Genes, Environments and Interactions

José M Álvarez-Castro

Genes, Environments and Interactions

Evolutionary and Quantitative Genetics Brought Up-to-date

Springer

José M Álvarez-Castro
Department of Education
University and Professional Training, Xunta de Galicia
Santiago de Compostela, Spain

Department of Statistics, Mathematical Analysis, and Optimization
University of Santiago de Compostela
Santiago de Compostela, Spain

ISBN 978-3-031-41158-8   ISBN 978-3-031-41159-5 (eBook)
https://doi.org/10.1007/978-3-031-41159-5

Cover image courtesy of Antonio Durán (creator) based on an idea by José M Álvarez-Castro.

This Springer imprint is published by the registered company Springer Nature Switzerland AG
The registered company address is: Gewerbestrasse 11, 6330 Cham, Switzerland

Paper in this product is recyclable.

Preface

This book is about the theory that formalizes the foundational concept of genetics—the way phenotypes are conditioned by underlying hereditary and environmental factors. The original setting of this theory provided a common framework for the study of Darwin's theory of evolution, Mendelian inheritance, and Galton's regression toward mediocrity, thus unlocking the takeoff of genetics as a scientific discipline. In recent times, limitations of this theory were pointed out in relation to its implementation in mapping experiments and other applications involving molecular data.

The mathematical expressions linking phenotypes to their underlying genotypes can be called genotype-to-phenotype (GP) maps—even in cases where not only differences in genotypes but also differences in environments are responsible for consistent phenotypic differences. That link is established through the so-called genetic (and environmental) effects, wherefore those mathematical developments are also called models of genetic (and environmental) effects. The two ways of referring to the expressions in question are considered synonyms throughout this book. The term GP map is used more often in relation to the original proposals—which established a sound theoretical link between genotypes and phenotypes. Then, as the expressions become more flexible, thus being able to mathematically model a broad scope of biological features, they are more often explicitly referred to as mathematical models (i.e., as models of genetic effects).

The focus of this book makes it necessarily profuse in mathematical developments. Nevertheless, its original motivation is the advent of a conceptual challenge. Thus, this book entails a conceptual and historical insight throughout, as its table of contents hopefully reflects. Furthermore, the theory provided is illustrated through consideration of several applied studies. A brief presentation of the table of contents actually follows and will in its turn be followed by some advice on how to read this book, including specifically some hints for readers without a strong mathematical background.

Annotated Table of Contents

The first chapter of this book reviews the obstacles that were faced to lay the foundations of the mathematical models satisfactorily describing general biological inheritance in terms of Mendelian factors, and how they were eventually overcome. That knowledge is valuable to better tackle subsequent challenges. In particular, a bundle of lessons are drawn for shedding light on how to upgrade the theory of biological inheritance to the gene-mapping era.

In the second chapter, the foundations of the mathematical theory of biological inheritance are reviewed. The expressions presented in this chapter provide adequate background for understanding the implementations of that theory that are developed in the following chapters. The specific challenges the original layout of the theory in question addressed are discussed, particularly with regard to glimpsing their limitations when used in alternative contexts.

The third chapter gathers several implementations of the original theory of biological inheritance, including explicitly considering epistasis, a convenient matrix notation, and the change-of-reference operation. In view of the theory presented, the distinction between analyses unaware and aware of a population context is clarified. Also, the convenience of a unified mathematical framework under which to develop a general GP map is highlighted.

A first step toward such a general framework—the natural and orthogonal interactions (NOIA) model—is given in the fourth chapter. In particular, models of genetic effects at the individual level are developed accounting for a wide range of genetic architectures. This chapter also delves into the meaning of the parameters of the models of genetic effects as effects of allele substitutions at the loci involved (and of shifts of environmental factors when those are also considered).

The fifth chapter extends the NOIA model from effects of allele substitutions at the individual level—i.e., performed from the reference of an individual genotype—toward averages of them over populations. The genetic architectures considered in the previous chapter are here taken to the population perspective accounting for a wide range of population features, just not including random associations between/among factors (loci and/or environments).

Those population features left aside in the previous chapter are precisely the focus of the sixth chapter. Since the developments accounting for these feature require some innovation of the mathematical approach employed, the resulting upgrade of NOIA is granted a label—ARNOIA (associations-resolved NOIA). With this, the NOIA framework achieves a high level of generality in what regards both the genetic architectures and the population features.

The seventh chapter deals with three peripheral theoretical implementations. First, expressions are provided that express the decomposition of the genetic variance with NOIA. Next, a statistical tool is described that maintains the properties of NOIA when estimating genetic effects from data with missing genotype information. And then an extension of NOIA also embracing average excesses is developed. The role of orthogonality is discussed throughout.

The eighth chapter gathers three illustrative applications. First, the maintenance of the multiallelic polymorphism of the human *ACP1* enzyme in a European population is explained. Next, Bateson-Dobzhansky-Müller incompatibilities are analyzed under departures from equilibrium frequencies. Last, a case of complex gene-environment interplay (concurrent interaction and correlation) is addressed. The convenience of an evolutionary quantitative genetics focus is highlighted.

The ninth chapter puts NOIA to test against the lessons drawn in the first chapter. Conceptual oscillations experienced about the pros and the cons of different research strategies in the recent history of quantitative genetics are revised along the way. As a theoretical proposal proven consequential for the different strategies, NOIA is shown to aid a consensus toward an enriched evolutionary quantitative genetics.

One of the lessons addressed in the ninth chapter urges to seriously valuing innovative proposals. For that reason, an addendum is devoted to reassuring the genetics community that, whether mathematically involved, NOIA is a necessary theory of genetic effects in the twenty-first century. In particular, an acid test is done by challenging NOIA with the principles of measurement theory. Along the way, the key concepts of the book are reviewed from different perspectives.

Tips on How to Read This Book

As mentioned above, the following tips are written particularly with readers who are not focused on mathematical expressions in mind. Yet, they may also be useful even to theoreticians. Indeed, any reader may prefer to address the details of the most mathematically involved passages only after having a clear conceptual picture in mind.

As a first consideration, Chaps. 1, 8, and 9, as well as the Addendum, are simply not mathematically involved and therefore no comment is necessary in this regard. Although Chap. 2 already is profuse in mathematical expressions, again no specific comments should be required here. The point is that Chap. 2 presents the theoretical foundations of quantitative genetics, which are familiar at least to a certain extent to most potential readers of this book. Thus, readers can decide to what extent to delve into the mathematical details of this chapter according to their specific needs.

Chapters 3 and 7 present several mathematical expressions that are not considered within the basic knowledge of quantitative genetics. In general, however, there are not many cumbersome expressions in these chapters. Thus, again, a routine reading strategy can be applied—to just pay as much attention to each of the equations as the reader feels like when going through it. Then, we are left with Chaps. 4–6 alone.

Indeed, the core of this book is the updating of the GP map to the current challenges in genetics research and it is presented in Chaps. 4–6 under the name of NOIA. These chapters thus compile theoretical work developed over a period of about 15 years (and published from 2007 to 2020) together with additional not previously published mathematical developments that complete it into a coherent, compact, and general theoretical framework. Indeed, many of the expressions already published are re-derived differently in this book, which thus provides them all into a common developmental structure. These chapters are therefore profuse in mathematical expressions that are not familiar to the general reader.

It is surely useful for the reader to know this in advance. Nevertheless, the key message under this heading is that these chapters have been written with the intention that the first section (the introduction) and the last one (the discussion, although not named that way) make sense on their own. This way, the reader has the possibility of reading them both first and, based on that, decide how much effort to put into each of the different sections in between. In short, a kind of friendly spoiler strategy is given as an option to read Chaps. 4–6.

That same strategy can also be applied at the level of the book itself. The reader can address the introduction (first chapter) and the aplication and the discussion (eight and ninth chapters, respectively) in a row. This way, the later reading of the intermediate Chaps. 2–7 may work easier at times, as it may become more obvious to percieve why some content is important to cover. In any case, the last chapters are correclty placed where they are—it should be of help to read them after having gone through the rest of the book even if they have already been viewed earlier.

Finally, leaving aside the complexity of the mathematics presented in this book, it is also considerably involved conceptually, as already mentioned. Figuring out the best ways to use of the GP map is tricky when different groups led by prestigious reserachers have held divergent views on this issue throughout the history of the field. This is to say, focusing on extracting the best of each historical proposal into a comprehensive view probably entails the most important challenge of this book.

Santiago de Compostela, Spain José M. Álvarez-Castro

Synopsis

Genetic effects are the nuclear concepts that gave rise to quantitative genetics and the evolutionary synthesis. They were formalized 35 years in advance of the discovery of the double helix of DNA, from which molecular genetics emerged. The theory of genetic and environmental effects is here revised and brought up-to-date about a century after the first models of genetic effects were developed and a generation after the advent of gene-mapping, thus providing several major advantages.

First, this book encompasses the widest range of simultaneously multiallelic and multilocus architectures, accounting for both autosomal and sex-linked loci with arbitrary interactions (whether dominance, gene–gene, gene–sex, parent-of-origin, or gene–environment interactions).

Second, it entails genetic effects as effects of allele substitutions both focusing on individual organisms and in a population context.

Third, in the latest case it fits to arbitrary departures from equilibrium frequencies (namely, departures from the Hardy-Weinberg equilibrium and from linkage equilibrium, and gene-environment correlations).

Fourth, it unifies the individual and the population approaches under the same framework, so that it easily enables transformations of genetic effects between different contexts (e.g., from the sample used in a mapping experiment into the putative original genotype from which a population diversified, or into a current population of interest).

And fifth, it does all the previous while sticking to the classical conceptualization of genetic effects and variance decomposition that made quantitative genetics and the evolutionary synthesis spring.

The concepts and theoretical developments here presented aid applications watering the field of evolutionary quantitative genetics—indeed, they are illustrated through built-in cases and real-data analyses—and they are corroborated based on the fundamentals of model development.

Acknowledgments

I was introduced into models of genetic effects in 2003 by Prof. Thomas Hansen when he was working as Assistant Professor at the Florida State University, USA. He is now working at the University of Oslo, Norway, from where he has sent me valuable suggestions on several chapters of this book. My conceptual understanding of quantitative genetics was shaped with his kind help and this book owes a significant debt to him.

I published some of the core theoretical results gathered in this book in co-authorship with Prof. Örjan Carlborg (Uppsala University, Sweden; original setting of NOIA), Prof. Rong-Cai Yang (University of Alberta, Canada; multiallelic NOIA), and Dr. Rosa Crujeiras (University of Santiago de Compostela, Spain; population models with departures from linkage equilibrium). Another of these results (biallelic gene–environment interactions without correlation between genes and environments) was developed by Dr. Jianzhong Ma (University of Texas, USA) and collaborators. I would also like to acknowledge occasional theoretical insight from Prof. Lars Rönnegård (Dalarna University, Sweden) and Dr. Andrés Prieto (University of A Coruña, Spain).

Many of the applications of the theory here presented, as well as the first bioinformatics tools implementing it, were possible through a long-lasting collaboration with Dr. Arnaud Le Rouzic, which started when we were postdoc colleagues at the Uppsala University, Sweden, in 2006. He is now working at the Paris-Saclay University, France. Beyond objective contributions to common publications, Dr. Arnaud Le Rouzic has afforded me countless priceless hours of fruitful discussion. During the writing of this book, I have been privileged to count on the friendly encouragement of Dr. Diddahally Govindaraju (Harvard University and Albert Einstein College, USA). He has also sent me insightful comments on several parts of the book. Jaime Imaz (AllGenetics, Spain) helped me spot several mistakes in the proofs shortly before publication.

Being married to a geneticist has certainly been of great help as well—just not as much as being married to an exceedingly generous, empathetic, and brilliant person,

Ania. I owe our son Galván so many hours, days, weeks... and I fear I will never be able to correspond his understanding. I deem my motivation to take on the challenge of this book to have been fueled by the self-confidence my parents and grandparents, Antonieta, Xabier, Kari and Baldomero, helped me develop as a child. It would be gratifying that what you are about to read shows that I have taken some advantage of all the help that was given to me.

Contents

Discovering the Genotype

1

Abstract

For some decades now, gene mapping has challenged the theory linking genotype and environment to phenotype—the mathematical description of the genotype-to-phenotype (GP) map, also accounting for underlying environmental factors. The present book takes up that challenge by providing a mathematical model of genetic effects that generalizes previous proposals and addresses the needs that arose particularly in the context of gene mapping. The idea of analyzing the influence of hereditary factors (genetic diversity) on phenotypes goes all the way back to Mendel in the nineteenth century, while Fisher established, during the first quarter of the twentieth century, the basis for applying Mendelian inheritance to the study of biometrics and gradual evolution. In this chapter, obstacles that severely hampered the foundation of quantitative and population genetics at that time are put on the table. Such reflection aims, on the one hand, to shed light on how to unlock current controversies about updating the GP map to the gene mapping era and, on the other hand, to pave the way for understanding the proposals conveyed in this book.

1.1 Introduction

It is common knowledge that genetic bases of phenotypes of interest can nowadays be inspected. Geneticists generally acknowledge that such bases putatively involve complex responses to varying environments. Naturally, one of the limiting steps for unveiling the genetic and environmental factors underlying a phenotype is to properly conceptualize and model every single way in which phenotypes may be shaped by genes and environments—we need to ensure that our mold does not suffer from limitations that may impose preset shapes on the object we intend to bare. In other words, gene mapping relies on adequate expressions of the phenotype in terms of underlying genotypes and environmental factors (e.g., Hansen and Wagner 2001;

J. M. Álvarez-Castro, *Genes, Environments and Interactions*,
https://doi.org/10.1007/978-3-031-41159-5_1

Yang 2004; Zeng et al. 2005; Álvarez-Castro and Carlborg 2007; Álvarez-Castro 2020).

Such expressions can thus be thought of as functions that assign genotypes (and environmental factors) to phenotypes. They are called genotype-to-phenotype (GP) maps, a term coined by Richard C Lewontin half a century ago (Lewontin 1974). Throughout this book, we shall refer to particular instances of GP maps as genetic architectures—which, again, may or may not involve different environmental factors (e.g., Álvarez-Castro 2016). The GP map is a core concept that condenses the original question that gave rise to genetics (see, e.g. Orgogozo et al. 2015). It is thus not surprising that many crucial matters in genetics, other than gene mapping, rely on adequate GP maps.

For example, mutational variances vary substantially for different genetic architectures (e.g., Jones et al. 2014). Also, tuned mathematical expressions connecting genotype and environment to phenotype are necessary for properly studying the effects of selection on phenotypes and, particularly, impediments to the action of selection due to, e.g., evolutionary plateaus or fitness local maxima traps (e.g., Goodnight 2015). The mathematical models in question are equally necessary for inspecting the limitations imposed in the reverse manner, i.e., how genetic architectures may and may not evolve—what is their evolvability—under different selection regimes (e.g., Carter et al. 2005). To mention just one more instance (which also uses the models in reverse), a fine-tuned enough GP may enable us to obtain fitness estimates from equilibrium population frequencies (Álvarez-Castro and Yang 2011).

Overall, it becomes crucial that a suitable theoretical formulation of the GP map is developed and used appropriately. In order to address that task, it is sensible to first examine the strengths and weaknesses of the theory currently in use. In this respect, it is crucial to revise how the different proposals developed in time, which will enable us to, on the one hand, understand the needs they met and, on the other hand, perceive the limitations imposed by the historical context under which they were carried out. Ultimately, we target a comprehensive theory that not only gathers all the advantages and none of the shortcomings of each of the previous proposals but also fills any remaining gaps. We shall thus conceptually dissect the state-of-the-art GP map in current genetics. In this chapter, in particular, we review several key events in the gestation of the basic mathematical theory of traits with Mendelian inheritance—i.e., of what we today can refer to as the seedling of the GP map.

1.2 Harnessing a Wealth of Experience

1.2.1 Two Lessons from Mendel

For the purpose of this book, it is neither necessary to have in mind a profuse overview on the history of genetics nor, particularly, of its deepest roots. Nevertheless, certain aspects of the origins of this scientific field may provide us with some useful clues to face the present-day challenges. Thus, we shall at least recall two consequential features of Mendel's foundational contribution. One of them deals with Mendel's outreach in his day, while the other one highlights a more specific aspect of his work.

As a general statement, Mendel's contribution entails a paradigmatic case of a remarkable breakthrough whose understanding and practical use by the scientific community suffered a significant delay. There is historical record supporting that the disclosure of Mendel's results was much further beyond an article in a low impact journal—or of its equivalent in 1866. Mendel not only bothered to expound his work and publish it in appropriate ways but also took the initiative of establishing a direct and long-lasting interchange of comments, questions, and materials with the community of researchers working in his field of study. Indeed, the precise understanding of Mendel's work for laying the foundations of genetics during the so-called rediscovery of Mendel's work—an event that, in a broad sense, Mendel himself predicted—was not based only on his publications and notebooks but also on his correspondence (see, e.g., Stern and Sherwood 1966).

Whether we think that we are able to understand in detail why a community focused on the questions Mendel addressed did not appropriately value his accessible and landmark achievement or not, it looks sensible to try to avoid similar errors when dealing with current questions and answers. Incidentally, we keep on focusing on matters very related conceptually with those Mendel addressed. Mendel inferred (by studying hybridization in plants) underlying hereditary factors of phenotypes, and one of our major current aims is to hone methodologies to dissect genetic and environmental factors underlying the variability observed in traits of interest. Disregarding contributions that overcome limitations in our capability to identify those factors implies to capriciously undermine our chances to deliver to our society solutions to critical problems in fields as vital and diverse as medicine, agriculture, livestock production, or conservation biology.

Now, it is known that if, in the aforementioned endeavor, we do not deal with interactions appropriately, we may not only be doomed to a much less efficient (or even misleading) use of the factors identified, but we could well also fail to identify many of the factors themselves (e.g., Carlborg and Haley 2004). Mendel already had to deal with such difficulty. Indeed, he could identify the hereditary factors of the phenotypes he studied only by assuming that the phenotypic ratios 3:1 he observed in his crossings were actually underlain by 1:2:1 ratios at the level of the hereditary factors. Such underlain ratios appeared in his phenotypic data masked as 3:1 due to an interaction called complete dominance.

Thus, the second and more specific point referred to above could be expressed by stating that interactions proved to be the clues to understanding crucial facts already at the very beginning of the study of biological inheritance. In other words, Mendel's experience already warns us that not paying enough attention to interactions comes at the risk of missing the actual explanatory mechanisms of the subject under study. In mapping experiments, in particular, the explanatory potential of interactions makes it worth devoting significant effort to pondering them before carrying on the analyses with additive effects alone. We could at this point feel tempted to restrain that advice only to what regards complete dominance—the interaction

Mendel found out—thus still disregarding the potential occurring of e.g. overdominance and epistasis. It may already seem whimsical to take for granted that one only interaction matters. Nevertheless, we shall discuss that temptation in more detail throughout this book.

1.2.2 One Lesson from Mendel's Rediscovery

Over three decades after its publication, the so-called rediscovery of Mendel's scientific legacy in 1900 was not free from controversy. On the contrary, it gave rise to a severe conflict between two divergent perspectives—those of Mendelians and biometricians. Mendelians became immediately impressed with the explanatory power of Mendel's contribution and focused on the study of evolution as a succession of abrupt steps. Biometricians, on the other side, advocated Darwinian gradual evolution and witnessed Mendel's results as a weapon in the hands of their opponents (Provine 1971).

There was a common denominator for the two opposing sides, but they eventually had to face the uncomfortable paradox of being equally wrong about it. One aspect of this common denominator was the reluctance to consider gradual evolution under the perspective of Mendelian inheritance. It has been pointed out that "if the Mendelians opposed selection and gradual change in some ways, they embraced them in others" (Stoltzfus and Cable 2014). Mendel himself had foreseen the potential of his findings to also analyze (Darwinian) continuous variation (Provine 1971). In any case, both the biometricians and the Mendelians in general lacked interest in analyzing gradual evolution by means of Mendelian inheritance (Provine 1971; Stoltzfus and Cable 2014), as further discussed below.

An even more patent aspect of their flawed common denominator was that both of them assumed that Mendelian inheritance would be incompatible with selection bringing a trait beyond its observed range (Stoltzfus and Cable 2014). The Mendelians stressed the role of mutations in overcoming preexisting phenotypic boundaries, while the biometricians thought that continuous (quantitative) variation could evolve under rules different from those found by Mendel when studying meristic (discrete) traits (Provine 1971; Stoltzfus and Cable 2014). Today, it is well established that the resulting phenotypes of a cross can lay outside the range defined by the parentals (i.e., there can occur transgressive segregation) not only due to recombination but also to underdominance, overdominance, and epistasis (e.g., Rieseberg et al. 2003; Álvarez-Castro et al. 2012).

The theory that enabled methodical analyses of gradual evolution of traits with Mendelian inheritance by expressing phenotypes "on the supposition of Mendelian inheritance" was published by Ronald Fisher almost two decades after the foundation of genetics (Fisher 1918). These years between 1900 and 1918 were described by Stoltzfus and Cable (2014) as "a period of dizzying experimental, conceptual and theoretical innovation." These authors also underscored that the conflict between Mendelians and biometricians during that period was based on a genuine "theoretical clash"—as opposed to a mere misunderstanding or the like.

Such dizzying innovation was developed by both parties during that period, in any case, in the context of explicit "personality clashes," as Provine (1971) profusely documented. Indeed, this author deems the effort directed to personality clashes (rather than just to a necessary, inspirational debate) to have delayed a consensus (and, thus, the resolution of the aforementioned theoretical clash) for about 15 years. Incidentally, this notice makes it difficult to avoid recalling that those 15 years came on top of the aforementioned more than three decades it had previously taken scientists to acknowledge the value of Mendel's experiments.

There is one single fact that probably illustrates this delay by itself. Already in 1902, Udny Yule (a biometrician) went beyond the ongoing personality clashes and provided solid mathematical developments supporting the compatibility between Mendelian inheritance and gradual evolution (Yule 1902)—in line with Mendel's intuition mentioned above. We can refer to these developments as an anticipation of Fisher's (1918) more complete theory. At the time, however, both parties turned a deaf ear to Yule's proposal (Provine 1971). Apparently, the tone of the discussion had already raised so loud that Yule's (1902) appeasing voice could not be heard. Just to mention one more fact about how convoluted the conflict was, let us also bring up here that the Biometric School had been founded on the pillars of the work of Francis Galton (1886), who nevertheless took sides in favor of the Mendelians—thus reluctant to gradual evolution, which had in its turn been postulated by his cousin, Charles Darwin (e.g., Provine 1971).

Fisher's (1918) theory set the foundations of the mathematical formulation of the GP map. It enabled geneticists to overcome harsh conflicts, and it has provided a basis to quantitative genetics analyses for the latest hundred years. In recent times, we can spot this theory again on the radar screen. Indeed, as mentioned above, the advent of mapping experiments has during the latest three decades challenged the way the effects of genetic and environmental factors are modeled, both conceptually and mathematically. In the next chapter of this book, we shall revise Fisher's (1918) theory as a first step to bring it up to date with the current uses it is to serve.

Reaching a consensus on how best to update and apply the GP map will not necessarily be an easy challenge. In particular, it would be desirable not to get stuck in the same kind of discrepancies as Mendelians and biometricians did. More to the point, it could be of help to us to learn how they in the end overcame those problems, as this could give us clues on how to find a consensus for our current challenge. Thus, in the following section, we shall have a brief look at the context in which gradual evolution based on Mendelian inheritance was generally accepted and investigated.

1.2.3 A Suitable Scenario for the Synthesis

Two kinds of contributions caused the context between the publication of Yule's (1902) paper and Fisher's (1918) to change significantly (as noted by Provine 1971). On the one hand, several experiments contradicting major postulates of the Mendelians were sequentially disclosed. First, William E Castle showed that selection could bring a continuous trait beyond the limits of the original phenotypic variation. Next, Herman Nilsson-Ehle and Edward M East found evidence of Mendelian basis for certain phenotypes with continuous variation. Lastly, Thomas Morgan and collaborators at his *Drosophila* laboratory ("the fly lab") eventually observed Mendelian traits with very small phenotypic variations. Overall, the idea that Mendelian inheritance naturally produces continuous variation susceptible of gradual evolution by natural selection gained strength, which raised an audience curious to better understand the underlying mechanisms.

On the other hand, two other theoretical works made Yule's (1902) seed to grow bigger. One was that of Wilhelm Weinberg, who derived expressions of correlations between close relatives under Mendelian inheritance in 1909 and 1910 that were consistent with quantitative data. And the other one came from Howard C. Warren in 1917, who also provided derivations in support of the compatibility between Mendelian inheritance and gradual Darwinian selection.

To be precise, it should be also noted that, although doubtlessly commendable, Weinberg's contribution probably provided little practical help to the acceptance of Fisher's (1918) work since, because of having been published in German, had very low impact. Indeed, the now-called Hardy-Weinberg principle was for many years named Hardy's law, even when Weinberg published his version of that principle, also in German, half a year in advance of the mathematician Godfrey H Hardy (from Cambridge, England).

Whichever the weights of the different stones, a suitable path was paved for Fisher (1918) to go beyond Yule's (1902) defense of the compatibility between Mendelian inheritance and gradual evolution towards a more complete and practical synthesis of Mendelism, Darwinism, and biometry. To that aim, he derived the key biometrical properties of a population under Mendelian inheritance. Such properties involved observable population phenotypic means and variances. In the words of Oscar Kempthorne (1968): "The problem that is addressed in the paper is to determine what genetical interpretation of data can be made when genotypes cannot be identified. Under this circumstance, the only observables are observations of related individuals, and the aim is to develop a theory, based on Mendelism, of the variability and covariability of these observables."

For developing such theory, Fisher (1918) specifically noted the need (and feasibility) to disentangle the effects on the phenotypic variance coming from different key sources—the difference between homozygotes (i.e., addictiveness), the departure of the heterozygotes from the midpoint between the homozygotes (i.e., dominance) and the environment. Incidentally, Yule (1902) had previously deemed the effects on phenotype of the latter two sources to be indistinguishable.

All in all, by showing that Mendelian inheritance could explain the observed phenotypic correlations between relatives, Fisher (1918) added up to eroding the assumptions that sustained the "theoretical clash" (and, therefore, also the "clash of personalities") between the Mendelians and the biometricians. Beyond that, he actually used Mendelian inheritance as a basis to develop a method for biometric analyses in practice. He developed expressions providing phenotypes in terms of separate additive and dominance effects of Mendelian factors and of environmental factors—he planted the seed for the GP map. We are currently in the need of using the resulting plant in a context Fisher could hardly foresee—the era of DNA sequencing and gene-mapping. In the next chapter we shall thus give a careful account of what Fisher's foundational GP map actually provided—and what it did not.

1.3 Drawing Lessons

It shall therefore be of great use throughout the reading of the present book to keep in mind at least four warnings related to the GP map that can be derived from contributions made even before the foundational GP map was developed. First, it is worth noting that severe avoidable delays in scientific progress have occurred since the very beginnings of the field of genetics and, particularly, in relation with the connection between genotype and phenotype. It is clear that a new proposal does not have to be immediately accepted just because it looks revolutionary to whatever extent, but it is also true that, more frequently than desired, new scientific proposals, and particularly challenging warnings, are not evaluated with the detail they deserve. This problem could in particular entail a burden today to keep the theory of genetic effects up-to-date with the possibilities offered by statistics, molecular genetics, and computational sciences.

Second, the broad explanatory power of interactions must always be kept in mind. Analogous to the previous point, it is clear that complex interactions must not blindly be assumed against simpler hypotheses. Indeed, it looks sensible to first inspect to what extent simpler hypothesis may suffice. More to the point, relatively simple models may be useful as predictive tools even when known not to be causative. Nevertheless, it should be kept in mind that interactions were crucial already to make the first move in the understanding of the connection between genotype and phenotype. Thus, albeit interactions may highly increase the complexity of a research project, experience tells us that reluctance to consider them, or even to consider them in depth, must be adequately justified, for it carries the risk of preventing any understanding of the phenomenon under study.

Third, personality clashes have also since early stages dulled our understanding of biological inheritance and the development of methods to study natural and artificial selection. In theory, to confront proposals is supposed to fuel scientific progress rather than hindering it—we could even say that science naturally works as a succession of controversies. And that is precisely why it is crucial not to take controversies personally. However, we probably have to assume that personal and group inertias and interests will always threaten the optimal functioning of the research activity. As a matter of fact, pointing out drawbacks in conventional procedures is often not only received with apathy, as already discussed in our first point above, but likely even provoke a prejudiced refusal. Now, history unmistakably teaches us how harmful it may get to be not to tame such disapproval within the realms of objectiveness. We must keep in mind that the extent to which the research community is able to deliver returns to society will strongly depend upon how scientific controversies are managed. Indeed, some great scientists are sometimes remembered for overreacting against breakthrough advances of their contemporary colleagues rather than for their own valuable contributions.

And fourth, especially in view of the above difficulties, it may be decisive to focus as much as possible on making the most of any findings that may help us broaden our scopes and build bridges between seemingly opposed perspectives. It would thus be extremely useful in our subject matter to sense and seize a favorable context for implementing new concepts and theoretical developments that enable a revision of the conventional line of work on the GP map, particularly in what regards interactions. Let us keep in mind that both experiments and preliminary theoretical studies pointed to Mendelian inheritance to be compatible with biometrical studies and gradual evolution when Fisher's (1918) theory of genetic effects was paid attention, accepted, and further developed. One century later, theoretical and empirical studies have revealed serious limitations to neutrally assess the high explanatory potential of interactions in practice. Thus, it seems only logical that all available upgrades of Fisher's theory that may help us overcome such limitations should be compiled and appraised. That is precisely what the present book is aimed to do.

References

Álvarez-Castro JM (2016) Genetic architecture. In: Wolf JB (ed) Encyclopedia of evolutionary biology. Oxford Academic Press, Oxford, pp 127–135

Álvarez-Castro JM (2020) Gene-environment interaction in the era of precision medicine - filling the potholes rather than starting to build a new road. Front Genet 11:921

Álvarez-Castro JM, Carlborg Ö (2007) A unified model for functional and statistical epistasis and its application in quantitative trait loci analysis. Genetics 176:1151–1167

Álvarez-Castro JM, Yang R-C (2011) Multiallelic models of genetic effects and variance decomposition in non-equilibrium populations. Genetica 139:1119–1134

Álvarez-Castro JM, Le Rouzic A, Andersson L, Siegel PB, Carlborg O (2012) Modelling of genetic interactions improves prediction of hybrid patterns--a case study in domestic fowl. Genet Res (Camb) 94:255–266

Carlborg Ö, Haley CS (2004) Epistasis: too often neglected in complex trait studies? Nat Rev Genet 5:618–625

Carter AJ, Hermisson J, Hansen TF (2005) The role of epistatic gene interactions in the response to selection and the evolution of evolvability. Theor Popul Biol 68:179–196

Fisher RA (1918) The correlation between relatives on the supposition of Mendelian inheritance. Trans Roy Soc Edinburgh 52:339–433

Galton F (1886) Regression towards mediocrity in hereditary stature. Anthrop Inst Great Britain and Ireland 15:246–263

Goodnight C (2015) Long-term selection experiments: epistasis and the response to selection. In: Moore JH, Williams SM (eds) Epistasis. Methods and protocols. Springer, New York, pp 1–18

Hansen TF, Wagner GP (2001) Modeling genetic architecture: a multilinear theory of gene interaction. Theor Popul Biol 59:61–86

Jones AG, Burger R, Arnold SJ (2014) Epistasis and natural selection shape the mutational architecture of complex traits. Nat Commun 5:3709

Kempthorne O (1968) The correlation between relatives on the supposition of Mendelian inheritance. Am J Hum Genet 20:402–403

Lewontin RC (1974) The genetic basis of evolutionary change. Columbia University Press, Columbia

Orgogozo V, Morizot B, Martin A (2015) The differential view of genotype-phenotype relationships. Front Genet 6:179

Provine WB (1971) The origins of theoretical population genetics. University of Chicago Press, Chicago

Rieseberg LH, Widmer A, Arntz AM, Burke JM (2003) The genetic architecture necessary for transgressive segregation is common in both natural and domesticated populations. Philos Trans R Soc Lond Ser B Biol Sci 358:1141–1147

Stern C, Sherwood ER (eds) (1966) The origin of genetics. A mendel source book. W. H. Freeman and Company, San Francisco

Stoltzfus A, Cable K (2014) Mendelian-mutationism: the forgotten evolutionary synthesis. J Hist Biol 47:501–546

Yang R-C (2004) Epistasis of quantitative trait loci under different gene action models. Genetics 167:1493–1505

Yule GU (1902) Mendel's laws and their probable relations to intra-racial heredity. New Phytol 1:193–207, 222–238

Zeng ZB, Wang T, Zou W (2005) Modeling quantitative trait Loci and interpretation of models. Genetics 169:1711–1725

The Primeval Theory of Genetic Effects 2

Abstract

The basics of Fisher's theory of genetic effects and variance decomposition on which population and quantitative genetics were established are here reviewed. The expressions explained in this chapter entail the foundations from which additional mathematical developments are proposed in following chapters. Indeed, Fisher's one-century-old theory needs to be upgraded to fulfill demands arisen due to research opportunities not even imaginable when it was mounted, which are today enabled by astounding advances made in several fields. Thus, beyond providing a comprehensive account of Fisher's theory, special attention is paid in this chapter to the questions that theory originally addressed as compared to the ones a genotype-to-phenotype (GP) map is expected to unravel nowadays. Thus, the present chapter points in particular to limitations of Fisher's theory to be overcome, which are commented in more detail in the following chapter.

2.1 Introduction

At the origins of quantitative and population genetics, Fisher (1918) formalized how phenotypes are conditioned by genotypes and environments, i.e., what was eventually called the genotype-to-phenotype (GP) map (Lewontin 1974). It is striking that, in general terms, this way of modeling persists today. But it is also essential to beware that, precisely because of that, we have carried over several of its original limitations. In order to overcome those limitations, it is necessary to first of all clearly recognize them.

In the previous chapter, we have put on the table a number of issues that hindered the development and acceptance of the GP map, thus doing our best in order to circumvent similar difficulties in the improvement of that foundational theory nowadays. In this chapter, we will review that theory. On the one hand, dissecting the aforementioned limitations is particularly convenient for being able to properly

J. M. Álvarez-Castro, *Genes, Environments and Interactions*,
https://doi.org/10.1007/978-3-031-41159-5_2

identify and value the need of further implementations. On the other hand, reviewing the basic theory is also necessary for bringing to mind the starting point from which further implementations—which are presented in the following chapters—were developed.

For describing the phenotypes in terms of Mendelian inheritance, Fisher (1918) defined mathematical parameters reflecting how substitutions of variants (alleles) within the Mendelian factors (loci) affect the phenotypes. Those effects of allele substitutions are called genetic effects. Thus, we can now refer to Fisher's (1918) theory both as a GP map and as a model of genetic effects. We shall use the second option when we want to enhance that, as mathematical developments, these models are flexible to formally describe the effects of allele substitutions under different conceptual perspectives—a nuclear issue throughout this book.

In any case, for efficiently spotting the current limitations of Fisher's (1918) theory of genetic effects, we must pay special attention to the uses we intend to put it today, as compared to those for which it was originally mounted. For instance, both molecular biology, biostatistics, and computational power flourished exuberantly during the second half of the twentieth century and conveniently crystallized into a number of gene-mapping techniques. We shall thus not only present the developments on which implementations will eventually be proposed but also, and not least, which questions they were originally intended to answer and which were not. Hopefully, it will then become easier to infer which amends and improvements are to be made in this theory and how.

2.2 Fisher's Starting Point

Fisher (1918) begins his developments with the following sentence: "Let us suppose that the difference caused by a single Mendelian factor is represented in its three phases by the difference of the quantities a, d, $-a$, and that these phases exist in any population with relative frequency P, 2Q, R, where P+2Q+R = 1." Let us in our turn begin by looking at this simple starting point closely.

In the first place, Fisher (1918) focuses on "the difference caused by a single Mendelian factor," due to its different "phases," which, incidentally he further on refers to also as the "somatic" outcome of that "factor." In any case, that is currently known as the set of genotypic values of a diploid locus—i.e., the expected phenotypes of each of the possible genotypes at that locus. The way in which Fisher (1918) proposes the genotypic values to be "represented" responds to the need—mentioned right above—of describing Mendelian inheritance in a way that the interaction effect of dominance is set explicitly as a parameter of the model.

Dominance is the departure of the expected phenotype of the heterozygote from the midpoint between those of the homozygotes. Thus, without dominance, the heterozygote is at the same distance from each of the homozygotes—i.e., we can move between the expected phenotypes of the different genotypes by sequentially adding one only parameter. Consequently, Fisher (1918) proposed a scale in terms of a parameter that behaves additively and a separate parameter accounting for an

Table 2.1 Fisher's (1918) transformation of the genotypic values into an additive/dominance scale

Genotypic values	A_1A_1	A_1A_2	A_2A_2
Raw	G_{11}	G_{12}	G_{22}
Transformation	$G_{11} - \frac{G_{11}+G_{22}}{2}$	$G_{12} - \frac{G_{11}+G_{22}}{2}$	$G_{22} - \frac{G_{11}+G_{22}}{2}$
Transformed values	$G_{11}^{ad} = \frac{G_{11}-G_{22}}{2}$	$G_{12}^{ad} = G_{12} - \frac{G_{11}+G_{22}}{2}$	$G_{22}^{ad} = \frac{G_{22}-G_{11}}{2}$
Additive/dominance scale	$a = G_{11}^{ad}$	$d = G_{12}^{ad}$	$-a = G_{22}^{ad}$

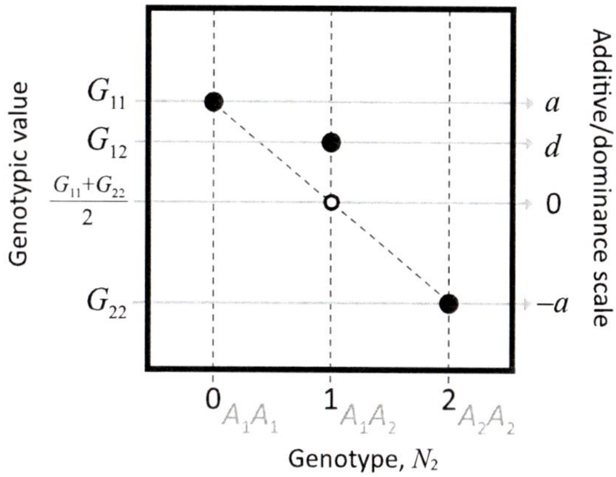

Fig. 2.1 Fisher's (1918) transformation of the genotypic values into an additive/dominance scale. The genotypes are represented in the horizontal axis of the graph by their content of allele A_2, N_2

interaction effect (dominance)—i.e., for a departure from the additive baseline. We shall hereafter refer to this scale as Fisher's additive/dominance scale.

2.2.1 Fisher's Additive/Dominance Scale

Let then the three genotypes be A_1A_1, A_1A_2, A_2A_2, with genotypic values G_{11}, G_{12}, G_{22}, respectively. Now, Fisher (1918) proposed to transform G_{11}, G_{12}, G_{22}, into a, d, $-a$, respectively, as shown both in Table 2.1 and in Fig. 2.1. With this transformation, the genotypic value of the heterozygote becomes the desired dominance parameter. What this transformation does is just to move the starting point of the measuring scale (which we can thus call the reference point) to the midpoint between the two homozygotes, $(G_{11} + G_{22})/2$, which thus becomes the reference of the transformed values. In other words, $a = G_{11}^{ad} = G_{11}–((G_{11} + G_{22})/2) = (G_{11}–G_{22})/2$, $d = G_{12}^{ad} = G_{12}–((G_{11} + G_{22})/2)$, and (as it could also be deduced from the definition of a) $-a = G_{22}^{ad} = G_{22}–((G_{11} + G_{22})/2)) = (G_{22}–G_{11})/2$.

It is easy to ascertain that the parameter d makes complete sense as a dominance parameter. On the one hand, d accounts for the presence/absence of dominance since it equals zero when the heterozygote, A_1A_2, lies right at the midpoint between the two homozygotes, A_1A_1 and A_2A_2 (i.e., when $G_{12} = ((G_{11} + G_{22})/2)$). On the other hand, d also accounts for the degree of dominance relative to the difference between the homozygotes. Indeed, a measures half the difference of the homozygotes and

d takes values between $-a$ and a when there is incomplete dominance, it equals either $-a$ or a when there is complete dominance (of hereditary factor A_1 or A_2, respectively), and it becomes lower than $-a$ or greater than a when there is underdominance or overdominance, respectively.

Thus, the transformation reaches the goal of expressing the genotypic values, G_{ij}, in a way that makes the dominance interaction, d, explicit while the remaining parameter, a, describes how the genetic system would behave (additively) without dominance. It is also noteworthy that the transformation implies no restriction since it can be done from any set of genotypic values, particularly including, in the words of Fisher (1918), the case when the heterozygote takes "any value between those of the dominant and the recessive, or even outside this range."

Such generality is not trivial. Comstock and Robinson (1948), for instance, kept the reference point of their scale at the same point as Fisher (1918) did but preferred the dominance deviation to be expressed as a ratio of the difference between the two homozygotes. Thus, instead of transforming the genotypic values into a, d, and $-a$, they transformed them into a, ad, and $-a$, respectively (although they actually modified the notation as well by using u instead of a, and a instead of d). In any case, Comstock and Robinson's (1948) parameterization is not completely general since it cannot account for the case of a heterozygote with a phenotype different from that of two phenotypically equal homozygotes. Indeed, $G_{11} = G_{22}$ implies that there is no additive effect (i.e., $a = 0$ since $a = -a$), which in its turn implies that the heterozygote cannot either differ from the homozygotes (since $ad = 0$) and thus, necessarily, $G_{11} = G_{12} = G_{22}$.

Nevertheless, generality may be kept in several other ways—it is possible to develop other transformations fulfilling the same task as Fisher's (1918). We could, for instance, just change Fisher's reference point to G_{11}, leading to the genotypic values to transform into 0, $a + d$, and $2a$. In fact, we shall in the following chapters analyze the convenience of this option in terms of its biological interpretation. At this point, however, we shall just try to understand why Fisher made that particular choice amongst other possible options.

To do so, the aforementioned attention call of Kempthorne (1968) about the "observables" proves very useful. Indeed, Fisher knew that it was necessary to make dominance explicit when expressing the genotypic values even if "genotypes cannot [i.e., they could not] be identified" because "the aim is [i.e., was] to develop a theory based on Mendelism, of the variability and covariability of these observables," that is, of the "observations of related individuals." In other words, why to bother in working out a tuned biological interpretation for parameters he could not even dream of observing? In fact, we now know that several decades would be needed for the molecular basis of Mendel's hereditary factors to be elucidated—through the experiments by Avery et al. (1944) and Hershey and Chase (1952). In that context, in addition to taking into account that the provided theory would be difficult enough to grasp for peers, putting the weight in simplicity at the expense of biological meaningfulness seems understandable.

Overall, the first particular task Fisher accomplished in his foundational publication was to find an as simple as possible set of parameters to express the genotypic values using an explicit term accounting for dominance, *d*. That is, he simply needed a yardstick, with no worries on its biological meaning. The solution he found is indeed simple since it expresses the genotypic values of the homozygotes in terms of the additive (i.e., unaware of dominance) contribution alone, *a*, and that of the heterozygote in terms of the dominance contribution alone, *d*.

2.2.2 Three Remarks on Fisher's Additive/Dominance Scale

2.2.2.1 Polarity

As noted above, even though Fisher (1918) was satisfied with his choice of a dominance scale for representing the genotypic values, he pointed out that it implied a certain "polarity" (of one "pure phase" over the other). More precisely, he developed his representation while naturally assuming that polarity to always exist but then acknowledged that this might not be the case. This can be seen by zooming out from our previous focus on his words about overdominance, further revealing that "[t]he heterozygote is from first assumed to have any value between those of the dominant and the recessive, or even outside this range, which terms therefore lose their polarity, and become merely the means of distinguishing one pure phase from the other."

From what we have just commented at the end of the previous section, no surprise about Fisher finding it useful to focus on "merely the means of distinguishing one pure phase from the other." But our point here is about the prejudice of a clean polarity caused by each Mendelian factor, which overdominance would already distort. Such prejudices may become fatal when developing theoretical models. In fact, special attention shall have to be paid for imposing no polarity on the Mendelian factors when laying a foundation for a general GP map in the following chapters of this book.

Further, it is at this point already possible to derive that, in general, interactions may not only "distort" to some extent a clean "polarity" but actually even reverse it. This can happen, for example, when the environments at which each of two different alleles is favored are different, leading to a kind of gene-environment interaction that reverses the order of the genotypic values. Just to mention a realistic general case, there may well be two different temperature ranges under which two different alleles reverse their relative performance. A similar situation occurs when a reversal is caused by a change in the genetic background, rather than in the environment—a relevant genetic mechanism known as sign epistasis (Weinreich et al., 2005).

Incidentally, Fisher (1918) assumed from the very beginning that epistasis could occur naturally, as dominance was assumed to do in the wake of the rediscovery of Mendel scientific legacy. He even bothered in elaborating an example to illustrate that epistasis could be widespread and devoted a section of his foundational article to mathematical developments with epistasis. Fisher's concern on epistasis was analogous to that on dominance and makes it clear that he assembled his theory based on the background concept of allele substitutions, which we address in our second remark.

2.2.2.2 Allele Substitutions

Fisher (1918) supported his choice of the additive/dominance scale by asking the reader to "consider what is involved in taking *a*, *d*, –*a* as representing the three phases of a factor. Genetically the heterozygote is intermediate between the dominant and the recessive, somatically it differs from their mean by *d*. [. . .] Somatically the steps are of different importance [. . .]. We may say that the somatic effects of identical genetic changes are not additive." That is, the same allele substitutions (the "steps") cause effects of different magnitude when performed over different genotypes, due to dominance. The same holds for epistasis: "A similar deviation from the addition of superimposed effects may occur between different Mendelian factors. We may use the term Epistacy to describe such deviation which although potentially more complicated, has similar statistical effects to dominance."

Thus, Fisher (1918) developed the additive/dominance scale consciously thinking of "allele substitutions [. . .] performed over [. . .] genotypes," i.e., at the individual level. However, we must recall once again that the particular epistatic effects of the genetic basis of quantitative traits could in general not be at the time considered as "observables," sensu Kempthorne (1968). Fisher's focus went indeed on the "statistical effects," caused on phenotypes at the population level by the individual-level allele substitutions. Overall, the additive/dominance scale is a yardstick reflecting allele substitutions at the individual level and it is designed for shedding light on the interpretation of the effects of gene interactions at the population level.

2.2.2.3 Avoid Confusion Between the Individual and the Population

Whichever side of the coin Fisher (1918) focused on, he gave a crystal clear warning about how easy and dangerous it was to confuse them. He particularly warned about certain "loose phrases" made at the time, "which obscure the essential distinction between the individual and the population [and therefore] should be avoided." This third remark of ours on Fisher's additive/dominance scale makes it evident that it was specifically designed to reflect properties at the individual level. Over a century later, we unfortunately may well interpret Fisher's warning as a premonition, as further justified below. A nuclear feature of the work presented in this book is precisely to provide novel theoretical tools finally enabling researchers to integrate both the individual and the population genetic perspectives in their analyses while staying true to the clear-cut conceptual distinction of both kinds of parameters.

Individual and population parameters are indeed associated to different practical implications. Because of that, researchers shall focus on one or on the other type of parameters in accordance with the issues they need to address. We shall hereafter explore how the GP can be expressed in terms of parameters accounting for population properties. We start from the one-locus population analysis and continue with some insight on multiple loci. In Chap. 3, we shall further develop on individual approaches associated to convenient biological interpretations and how they relate to the population approaches.

2.3 Fisher's One-Locus Population Analysis

Using Fisher's (1918) own words from the first two sentences of his foundational paper, the goal of that publication lays intrinsically at the population level and can be stated to be to "ascertain the biometrical properties of a population," particularly "in accordance with the Mendelian scheme of inheritance." To that aim, he firstly looked for an adequate scale providing (as discussed right above) "merely the means of distinguishing one pure phase from the other"—i.e., at the level of the genotypes, thus still regardless of the population level—to then being able to track how the departures from additivity (dominance, epistasis) at the level of the genotypes would cause a "deviation from the addition of superimposed effects" at the population level. Such population-level analysis shall be discussed hereafter.

The way in which Fisher originally presented his developments was later on criticized and improved. He provided little explanation about the basic theoretical setting and dealt also in an untimely manner with a high number of more specific situations. For instance, Kempthorne (1968) stated that Fisher's seminal work "constituted such a great advance on the thought in the area," "[t]he general methodology of the paper is very interesting but is not, I think, to be recommended for a first exposure to the subject," and, overall, "[t]he paper is difficult to read, not only for the sophistication of ideas, but also because it abounds in approximations, which are described only briefly, if at all."

Kempthorne himself reworked Fisher's theory in a more amenable manner, extended it, and appended historical notes on simpler results previously obtained by other researchers (Kempthorne, 1955a, 1957). There is also a rather popular and profuse commentary of Fisher's theory by Moran and Smith (1966)—several errata were found in this publication, compiled by, e.g., James J Lee.[1] The publication by Kempthorne (1968) cited above several times is actually a review of the book by Moran and Smith (1966).

In any case, the population analysis presented hereafter has been tailored specifically to what is required for understanding the theory developed in the subsequent chapters of this book. We still fit to the one-locus two-allele scheme introduced in Table 2.1. Table 2.2 gathers the essential parameters of the population analyses.

From the first row below the heading of Table 2.2, the population size, N, can be expressed as the sum of the individuals of each genotypic class, $N = N_{11} + N_{12} + N_{22}$, and the genotypic frequencies are $p_{11} = N_{11}/N$, $p_{12} = N_{12}/N$ and $p_{22} = N_{22}/N$. The number of alleles A_1 in the population is $N_1 = 2\,N_{11} + N_{12}$ and, likewise, that of alleles A_2 is $N_2 = 2\,N_{22} + N_{12}$. Then, $N_1 + N_2 = 2\,N$ and the allele frequencies, in the second row of the table, are $p_1 = N_1/2N = p_{11} + ½p_{12}$ and $p_2 = N_2/2N = p_{11} + ½p_{12}$.

Note that the notation of the population frequencies (with p_i being the frequency of allele A_i) is different from that of Fisher (1918) shown above (with genotypic frequencies of P, 2Q, and R). In general, our notation fits to current standards rather than to historical uses. Although Fisher (1918) carried out some more general

[1] https://jamesjlee.altervista.org/wp-content/uploads/2019/06/moran_and_smith_errata.pdf

Table 2.2 Essential parameters and variables of the population analysis

	A_1	A_2	A_1A_1	A_1A_2	A_2A_2	Variance
Population numbers	N_1	N_2	N_{11}	N_{12}	N_{22}	–
Population frequencies	p_1	p_2	p_{11}	p_{12}	p_{22}	–
Phenotypic level	–	–	P_k^{11}	P_k^{12}	P_k^{22}	V_P
Genetic level	–	–	G_{11}	G_{12}	G_{22}	V_G
Additive level	α_1	α_2	α_{11}	α_{12}	α_{22}	V_A
Dominance level	–	–	δ_{11}	δ_{12}	δ_{22}	V_D
Environmental level	–	–	e_k^{11}	e_k^{12}	e_k^{22}	V_E

developments, at this stage, we shall restrict our scope to the Hardy-Weinberg case, $p_{11} = p_1^2$, $p_{12} = 2p_1p_2$, and $p_{22} = p_2^2$, thus being able to raise the required issues in a more straightforward manner.

We shall first derive the essential expressions step by step through the different levels of the population analysis, which in fact follow the sequence of the remaining rows in Table 2.2 (the last five rows of that table). Within each level, we shall derive expressions for the corresponding effects and for their variance. In subsequent sections, we shall further comment on their biological meaning, interconnections, implications, applications, and limitations.

2.3.1 Phenotypic Level

As shown in Table 2.2, the phenotypes of the individuals of each genotypic class are P_k^{ij}, $k = 1, \ldots, N_{ij}$, $ij = 11, 12, 22$. The mean phenotype is the average of all of them regardless of its genotype and can thus be expressed in terms of each individual phenotype, P_n, $n = 1, \ldots, N$, as:

$$\mu = \left(\sum_{ij}\sum_{k=1}^{N_{ij}} P_k^{ij}\right)/N = \left(\sum_{n=1}^{N} P_n\right)/N. \tag{2.1}$$

From this expression, the phenotypic (also called total) variance can be straightforwardly obtained as:

$$V_{\mathrm{P}} = \left(\sum_{n=1}^{N} (P_n - \mu)^2\right)/N. \tag{2.2}$$

2.3.2 Genotypic Level

The usefulness of the aforementioned notation for distinguishing the phenotypes belonging to each genotypic class becomes clear when it comes to the genotypic level. Indeed, the three raw genotypic values (Tables 2.1 and 2.2) are defined as the average phenotype of each of the three genotypic classes, i.e.:

$$G_{ij}=\left(\sum\nolimits_{k=1}^{N_{ij}}P_k^{ij}\right)/N_{ij}, ij=11,12,22. \tag{2.3}$$

The population mean can also be obtained in terms of the genotypic values. Indeed, starting from Expression 2.1, we can easily derive:

$$\mu=\frac{\sum_{n=1}^{N}P_n}{N}=\frac{\sum_{ij}\sum_{k=1}^{N_{ij}}P_k^{ij}}{N}=\sum\nolimits_{ij}\frac{N_{ij}}{N}\frac{\sum_{k=1}^{N_{ij}}P_k^{ij}}{N_{ij}}=\sum\nolimits_{ij}p_{ij}G_{ij}=p_1^2G_{11}$$
$$+2p_1p_2G_{12}+p_2^2G_{22}.$$

Thus, the weighted (by the genotypic frequencies) average of the genotypic values is the population mean phenotype. Then the genotypic variance, V_G, is defined as the variance of the genotypic values and it can thus be derived from Expressions 2.1 and 2.3 as $V_G=\sum\limits_{ij}p_{ij}\left(G_{ij}-\mu\right)^2=p_1^2\ (G_{11}-\mu)^2+2p_1p_2(G_{12}-\mu)^2+p_2^2\ (G_{22}-\mu)^2$.

Nevertheless, we have stated above the purpose of expressing the population properties in terms of Fisher's additive/dominance scale (Table 2.1 and Fig. 2.1). To that aim, first we obtain the population mean under Fisher's additive/dominance scale as $\tilde{\mu}=\sum\limits_{ij}p_{ij}G_{ij}^{ad}=p_1^2\ a+2p_1p_2d+p_2^2\ (-a)$, leading to:

$$\tilde{\mu}=a(p_1-p_2)+2dp_1p_2. \tag{2.4}$$

It is easy to derive that Expressions 2.1 and 2.4 differ indeed by the additive/dominance scale transformation as:

$$\tilde{\mu}=\sum\nolimits_{ij}p_{ij}G_{ij}^{ad}=\sum\nolimits_{ij}p_{ij}\left(G_{ij}-\frac{G_{11}+G_{22}}{2}\right)=\left(\sum\nolimits_{ij}p_{ij}G_{ij}\right)-\frac{G_{11}+G_{22}}{2}=\mu$$
$$-\frac{G_{11}+G_{22}}{2}.$$

Second, we derive the mean-centered genotypic values, $\tilde{G}_{ij}=G_{ij}^{ad}-\tilde{\mu}$, as $\tilde{G}_{11}=G_{11}^{ad}-\tilde{\mu}=a-\tilde{\mu}$, $\tilde{G}_{12}=G_{12}^{ad}-\tilde{\mu}=d-\tilde{\mu}$, and $\tilde{G}_{22}=G_{22}^{ad}-\tilde{\mu}=-a-\tilde{\mu}$, for genotypes A_1A_1, A_1A_2, and A_2A_2, respectively, leading to:

$$\tilde{G}_{11}=2p_2(a-dp_1),\tilde{G}_{12}=a(p_2-p_1)+d(1-2p_1p_2),\tilde{G}_{22}=-2p_1(a+dp_2). \tag{2.5}$$

It is easy to verify that this adjustment of the mean is not affected by the scale in which it is performed, as:

$$\tilde{G}_{ij}=G_{ij}^{ad}-\tilde{\mu}=\left(G_{ij}-\frac{G_{11}+G_{22}}{2}\right)-\left(\mu-\frac{G_{11}+G_{22}}{2}\right)=G_{ij}-\mu \tag{2.6}$$

And finally, it is now straightforward—from Expression 2.5—to obtain an expression of the genotypic variance, V_G, in terms of a and d as $\sum\limits_{ij}p_{ij}\tilde{G}_{ij}^2=p_1^2$

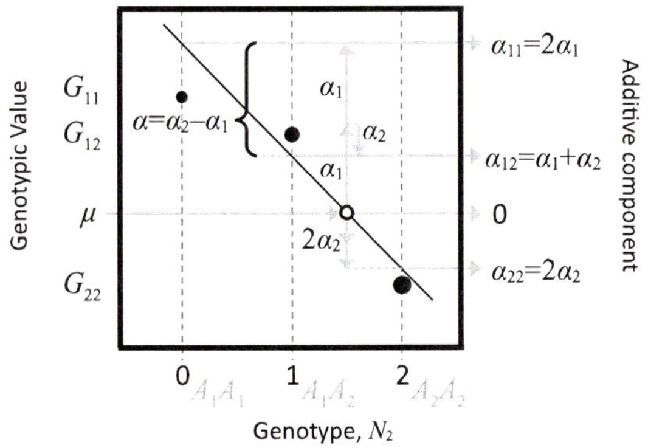

Fig. 2.2 Graphical interpretation of Fisher's (1918) additive population parameters—average effects of allele substitutions, α_i, $I = 1, 2$, and breeding values, α_{ij}, $ij = 11, 12, 22$. The genotypes are represented in the horizontal axis of the graph by their content of allele A_2, N_2. The black solid line is the regression (linear approximation) to the genotypic values

$(2p_2(a - dp_1))^2 + 2p_1p_2(a(p_2 - p_1) + d(1 - 2p_1p_2))^2 + p_2^2\,(-2p_1(a + dp_2))^2$. This can be worked out with some basic algebra to obtain the conventional expression (e.g. Falconer and Mackay 1996):

$$V_G = 2p_1p_2(a + d(p_2 - p_1))^2 + (2dp_1p_2). \tag{2.7}$$

2.3.3 Additive Level

The additive components of the genotypes A_1A_1, A_1A_2, and A_2A_2 are α_{11}, α_{12}, and α_{22}, respectively (fifth row below the heading of Table 2.2). They are defined as the values predicted by the least-squares linear regression to the genotypic values expressed as deviations from the mean phenotype, as illustrated in Fig. 2.2. Fisher referred to them as the "representative measures" (which, as opposed to the "real" ones, are represented by the linear regression). They are currently called breeding values because the values they take predict the average phenotype of the offspring of the individuals of the corresponding genotype, under random mating—i.e., they predict how the individuals of the different genotypes perform as breeders in the population under random mating. They behave additively as going from one to the next entails adding the slope of the linear regression. The variance of the breeding values is called the additive variance, V_A.

The breeding values are often expressed in terms of the average effects of allele substitutions, α_1 and α_2 (also included in the fifth row below the heading of Table 2.2). As shown by, e.g., Falconer and MacKay (1996), these can be inferred as the average phenotypes of the offspring produced by each of the alleles under random mating (which are $p_1a + p_2d$ and $p_1d - p_2a$ for alleles A_1 and A_2, respectively) minus the population average in Expression 2.4. After simplification they can be expressed as:

$$\alpha_1 = p_2(a + d(p_2 - p_1)) \text{ and } \alpha_2 = -p_1(a + d(p_2 - p_1)). \tag{2.8}$$

The effects of allele substitutions are here expressed in a way that their weighted average is zero.

The measure of the average effect of an allele substitution is conceptually tied to the selection process since selecting phenotypes will lead to genetic change as long as allele substitutions have on average an effect on the mean phenotype. It is thus not surprising that the performance of the individuals as breeders (or, alternatively, the distance between them) can be computed, as mentioned right above, from the average effects of allele substitutions.

Indeed, following, e.g., Falconer and MacKay (1996), [2] the average effects of allele substitutions in Expression 2.8 provide both the slope of the aforementioned linear regression—known as the additive effect of the gene at the population, α—and the breeding values, α_{11}, α_{12}, and α_{22}, as (see also Fig. 2.2):

$$\alpha = \alpha_2 - \alpha_1, \alpha_{ij} = \alpha_i + \alpha_j, ij = 11, 12, 22. \tag{2.9}$$

Then, using Expression 2.8, the slope of the regression in Expression 2.9 can trivially be expanded as:

$$\alpha = -(a + d(p_2 - p_1)), \alpha_{11} = -2p_2\alpha, \alpha_{12} = (p_1 - p_2)\alpha, \alpha_{22} = 2p_1\alpha. \tag{2.10}$$

The action of selection on the phenotype is more efficient in modifying the allele frequencies as this slope (the additive effect of the gene, α) departs from zero. Analogous to the allele substitutions in Expression 2.8, the weighted average of the breeding values in Expression 2.10 is zero.

The additive variance can now be obtained as the variance of these breeding values as $\sum_{ij} p_{ij}\alpha_{ij}^2 = p_1^2\alpha_{11}^2 + 2p_1p_2\,\alpha_{12}^2 + p_2^2\alpha_{22}^2$, or, using a shortcut, as the variance of the average effects of allele substitutions in Expression 2.8, as $\sum_i p_i\alpha_i^2 = p_1\,\alpha_1^2 + p_2\,\alpha_2^2$, which after simplification can be written as:

$$V_A = 2p_1p_2\alpha^2 = 2p_1p_2(a + d(p_2 - p_1))^2. \tag{2.11}$$

[2] These authors opted for performing the regression to the genotypic values by sorting the genotypic values in terms of the allele content of A_1 (see Figure 7.2 in that book) instead of that of A_2 as in Figures 1 to 4 above and below. That choice leads to some differences between Expressions 2.9 and 2.10, on the one hand, and the corresponding expressions by those authors, on the other hand.

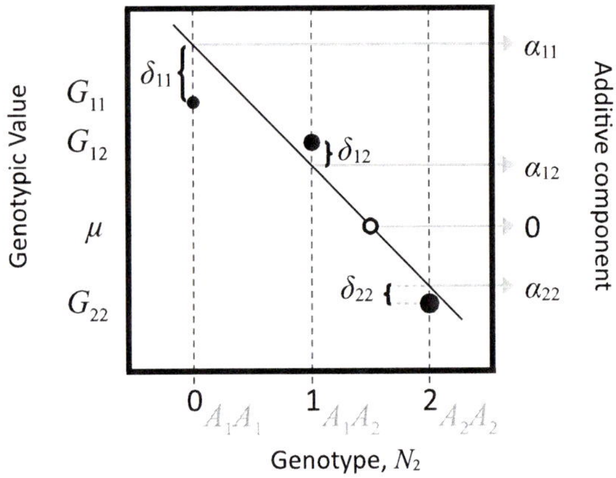

Fig. 2.3 Graphical interpretation of Fisher's (1918) dominance population parameters, δ_{ij}, $ij = 11, 12, 22$, relative to the additive population components. The genotypes are represented in the horizontal axis of the graph by their content of allele A_2, N_2. The black solid line is the regression (linear approximation) to the genotypic values

2.3.4 Dominance Level

As shown in Fig. 2.3 (see also sixth row below the heading of Table 2.2), the dominance component of each genotype is defined as the departure of the value predicted by the linear regression for that genotype and its genotypic value—the difference between the "representative measures" and the "real" ones. As departures of predictions caused by dominance, they are called dominance deviations. They can be derived directly from their definition as:

$$\delta_{ij} = \alpha_{ij} - \tilde{G}_{ij}. \tag{2.12}$$

Thus, they can be expresses using Expressions 2.5 and 2.10 as:

$$\delta_{11} = -2p_2^2 d, \delta_{12} = 2p_1p_2 d, \text{and } \delta_{22} = -2p_1^2 d. \tag{2.13}$$

Then the dominance variance, V_D, is just the variance of those values, $\sum_{ij} p_{ij}\delta_{ij}^2 = p_1^2\delta_{11}^2 + 2p_1p_2\,\delta_{12}^2 + p_2^2\delta_{22}^2$, which can be expressed as:

$$V_D = (2dp_1p_2)^2. \tag{2.14}$$

The expressions derived above for the genetic, additive, and dominance levels in terms of the additive/dominance scale—between Expressions 2.5 and 2.14—are gathered in Table 2.3.

2.3.5 Environmental Level

For defining the environmental component, we need to consider separately the phenotypes of the individuals of each genotypic class with the notation in

Table 2.3 Summary of expressions of variables of the genetic, additive and dominance levels in terms of a and d (Expressions 2.5–2.14, Table 2.1 and Fig. 2.1). All values are scaled by the phenotype mean. The weighted (by the genotypic frequencies) mean of each set of allele or genotype values is zero. The additive values of the genotypes are related to those of the alleles as $\alpha_{ij} = \alpha_i + \alpha_j$. The dominance values of the genotypes are defined as $\delta_{ij} = \tilde{G}_{ij} - \alpha_{ij}$. Additionally, the slope of the least-squares linear regression to the genotypic values is $\alpha = -(a + d(p_2 - p_1))$

	A_1	A_2	A_1A_1	A_1A_2	A_2A_2	Variance
Genotypic level	–	–	$\tilde{G}_{11} = 2p_2(a - p_1d)$	$\tilde{G}_{12} = a(p_2 - p_1) + d(1 - 2p_1p_2)$	$\tilde{G}_{22} = -2p_1(a + dp_2)$	$V_G = 2p_1p_2(a + d(p_2 - p_1))^2 + (2dp_1p_2)^2$
Additive level	$\alpha_1 = p_2(a + d(p_2 - p_1))$	$\alpha_2 = -p_1(a + d(p_2 - p_1))$	$\alpha_{11} = 2p_2(a + d(p_2 - p_1))$	$\alpha_{12} = (p_2 - p_1)(a + d(p_2 - p_1))$	$\alpha_{22} = -2p_1(a + d(p_2 - p_1))$	$V_A = 2p_1p_2(a + d(p_2 - p_1))^2$
Dominance level	–	–	$\delta_{11} = -2\,p_2^2d$	$\delta_{12} = 2p_1p_2d$	$\delta_{22} = -2\,p_1^2\,d.$	$V_D = (2dp_1p_2)^2$

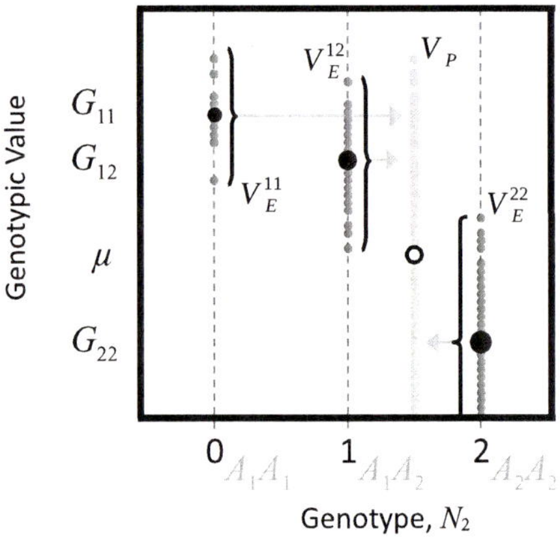

Fig. 2.4 Graphical interpretation of Fisher's (1918) environmental variance. The genotypes are represented in the horizontal axis of the graph by their content of allele A_2, N_2

Table 2.2 (third row below the heading), as illustrated in Fig. 2.4. Indeed, we have already seen above that when pulling all individuals together, as indicated by the arrows in that figure, their mean is the phenotypic mean of the population, μ, and their variance is the phenotypic variance, V_P, shown in Expressions 2.1 and 2.2, respectively. Now we are interested in where each of those sets of phenotypes is centered attending to the genotypes of the individuals, that is, on the genotypic values defined in Expression 2.3. This way we can define the environmental deviations (see the last row of Table 2.2) as:

$$e_l^{ij} = P_l^{ij} - G_{ij}, ij = 11, 12, 22, l = 1, \ldots, N_{ij}. \tag{2.15}$$

The genotype-wise environmental variances marked in Fig. 2.3 are the variances of each of those sets of phenotypes around their corresponding genotypic values, i.e., the variance of each set of environmental deviations:

$$V_{\mathrm{E}}^{ij} = \left(\sum\nolimits_{k=1}^{N_{ij}} \left(e_k^{ij}\right)^2\right) / N_{ij}, ij = 11, 12, 22, \tag{2.16}$$

whereas the (population-wise) environmental variance is the variance of the whole set of environmental deviations:

$$V_{\mathrm{E}} = \left(\sum\nolimits_{ij} \sum\nolimits_{k=1}^{N_{ij}} \left(e_k^{ij}\right)^2\right) / N. \tag{2.17}$$

Note that whether computing the variance of sets of environmental deviations or of all of them together, each environmental parameter remains a deviation from its corresponding genotypic value, rather than from the mean of the whole population. Otherwise, the computations would provide the phenotypic variance instead, as commented right above. Thus, Expressions 2.16 and 2.17 use the same parameters, centered on the genotypic values, and they actually are interconnected. Indeed, starting from Expression 2.17, we can derive an expression of it in terms of Expression 2.16 as:

$$V_{\mathrm{E}} = \frac{\sum_{ij}\sum_{k=1}^{N_{ij}}\left(e_k^{ij}\right)^2}{N} = \sum_{ij}\frac{N_{ij}}{N}\frac{\sum_{k=1}^{N_{ij}}\left(e_k^{ij}\right)^2}{N_{ij}} =$$

$$\sum_{ij} p_{ij} V_{\mathrm{E}}^{ij} = p_1^2 V_{\mathrm{E}}^{11} + 2p_1p_2 V_{\mathrm{E}}^{12} + p_2^2 V_{\mathrm{E}}^{22}.$$

Thus, the (population-wise) environmental variance in Expression 2.17 is the weighted (by the genotypic frequencies) mean of the genotype-wise environmental variances in Expression 2.16.

2.3.6 Theoretical Implications and Biological Interpretations

2.3.6.1 Already Mentioned Insights

Some aspects of the biological insight of Fisher's population analysis have already been pointed out as it has been derived right above. The most important ones are summarized hereafter, while further insights are discussed under the following headings. First, the genotypic values are the phenotypes expected for each genotype and they have been reparameterized in terms of the distance between the genotypic values of the homozygotes (the additive value, a) and the distance between the genotypic value of the heterozygote and the midpoint of the genotypic values of the homozygotes (the dominance value, d). This is Fisher's additive/dominance scale (Expressions 2.3–2.6), which provides an explicit genetic effect accounting for dominance, d, at the individual level—i.e., regardless of any population frequencies.

Second, population properties of interest can be expressed in terms of Fisher's additive/dominance scale, including the extent to which the genetic composition influences the phenotypic variance in a population—the genotypic variance (Expression 2.7). That genotypic variance can in its turn be split into the additive and the dominance variances (Expressions 2.11, 2.14), i.e., the separate contributions of the additive and the dominance genetic effects at the individual level, respectively, to the phenotypic variance of a population. And third, the additive variance is defined as the variance of variables with key biological interpretations—the average effects of allele substitutions and the breeding values of the genotypes (Expressions 2.8–2.10).

2.3.6.2 Decomposition of the Phenotype

Expression 2.15 implicitly provides a decomposition of the phenotype of each individual in the population into a genetic and an environmental component, as:

$$P_k^{ij} = G_{ij} + e_l^{ij}, ij = 11, 12, 22, k = 1, \ldots, N_{ij}. \tag{2.18}$$

Further, Expression 2.6 entails a decomposition of the genotypic value into the population mean phenotype and the corrected genotypic value, which enables to rewrite Expression 2.18 as:

$$P_k^{ij} = \mu + \widetilde{G}_{ij} + e_l^{ij}, ij = 11, 12, 22, k = 1, \ldots, N_{ij}. \tag{2.19}$$

Either way, the phenotype of each individual is split into parameters accounting for the genetic (conditioned by Mendelian hereditary factors) and the environmental contributions.

Either expression is equally useful for the next step we need to take—extending the previous decomposition to the level of the variances. Since the variance of the population mean phenotype (which is equal for all individuals) is zero, and the variance of the raw and the corrected genotypic values is equal, provided that the three genotypic classes are subject to the same environmental conditions; it can be derived from both Expressions (2.18, 2.19) that:

$$V_{\mathrm{P}} = V_{\mathrm{G}} + V_{\mathrm{E}}. \tag{2.20}$$

This expression actually assumes no covariance between genotypes and environments: *cov*(G,E) = 0. Otherwise, that covariance has to be added in the term on the right-hand-side of the equality.

Using the decomposition in Expression 2.20, the relative influence of the genetic and the environmental components of a trait at a population is commonly expressed in terms of the proportion of genetic variance in the total variance, known as the broad sense heritability, H^2:

$$H^2 = V_{\mathrm{G}}/V_{\mathrm{P}}. \tag{2.21}$$

The notation is always squared, just making it explicit that the heritability is defined in terms of variances—as opposed to, e.g., standard deviations. A broad sense heritability of ½ stands for equal influence; the influence of the environmental component (meaning, the environmental variance) is higher when H^2 takes values between zero and ½, and the influence of the genetic component (meaning, the genetic variance) is higher for values of H^2 between ½ and one.

2.3.6.3 Reading the Components of the Phenotype

The simple decompositions in Expressions 2.18 and 2.19 are useful to train ourselves in relation with Fisher's warning on the distinction between the individual and the population. We may in particular consider whether the decompositions in Expressions 2.18 and 2.19 could be different for two individuals that share the same genotype and the same phenotype but belong to populations that are kept under the same environmental conditions but have different allele frequencies.

If we use Expression 2.18, the decomposition should remain the same in the two populations. Indeed, on the one hand, the set of genotypic values (the phenotype expected for each of the genotypes) should remain constant under the same conditions, and we are comparing individuals with equal genotypes. On the other hand, the previous forces the environmental component to also keep the same value, since we assumed individuals with equal phenotypes as well.

However, if we use Expression 2.19 instead, then the decomposition clearly depends on the population frequencies, since so does the population mean phenotype. The corrected (by the population mean phenotype) genotypic value, $\tilde{G}_{ij}$, depends here upon how close the population frequency lays relative to the raw genotypic value, G_{ij}, used in Expression 2.18. The environmental component should remain the same, as in the previous case, but its magnitude could now be either higher or lower than the genetic component depending on the value the latest one takes.

In summary, as opposed to Expression 2.18, Expression 2.19 makes the genetic component of the individuals to depend upon the departure of the phenotype expected by its genetic composition from the population mean phenotype. Disturbing as it could look at a first glance, Expression 2.19 reflects biometrical properties of the population and it is the one that shall serve our purposes better later on.

The decomposition of the phenotypic (also called total) variance in Expression 2.20 and, particularly, the broad sense heritability in Expression 2.21 deal with a very interesting "biometrical property" of the population since they entail a comparison of the genetic and the environmental influences on the trait at that population. To this regard, it should not be conceptually confusing that the population analysis in Expression 2.20 may come from either an expression with parameters entailing properties at the individual level—Expression 2.18—or an expression already entailing properties at the population level—Expression 2.19. Indeed, by computing variances, we are using the genotype frequencies of a particular population, thus providing an expression applicable to the trait in that population—hence, not only to the trait itself.

2.3.6.4 Decomposition of the Genotype

The additive and the dominance levels in Expressions 2.8–2.14 enable a decomposition of the genotypic values and the genetic variance and, therefore, to deepen the above decomposition of the phenotype and the phenotypic variance. From the definition of the dominance deviations in Expression 2.12, the genotypic values can be expressed in terms of additive and dominance components as:

$$G_{ij} = \mu + \widetilde{G}_{ij} = \mu + \alpha_{ij} + \delta_{ij}, ij = 11, 12, 22. \tag{2.22}$$

Analogous to the previous case, the decomposition at the level of the variances can be derived from the decomposition of the effects as:

$$V_{\mathrm{G}} = V_{\mathrm{A}} + V_{\mathrm{D}}. \tag{2.23}$$

It is at this point worth recalling that we have assumed Hardy-Weinberg proportions throughout and thus, in particular, for deriving the expressions for V_{A}

and V_D above—Expressions 2.11 and 2.14, respectively. By comparing these expressions with Expression 2.7, it now becomes clear that the later has been kept in a form that makes the decomposition in Expression 2.23 explicit.

Combining Expressions 2.20 and 2.23, we straightforwardly obtain the aforementioned deeper decomposition of the phenotype as:

$$V_P = V_A + V_D + V_E. \tag{2.24}$$

As we shall further explain below, the additive variance indicates the proportion of the total variance that can be harnessed by selection on the phenotype to modify the mean phenotype of the population in the next generation. Such proportion is commonly expressed in terms of the quotient of the additive and the total variance, known as the (narrow sense) heritability, h^2:

$$h^2 = V_A / V_P. \tag{2.25}$$

2.3.6.5 Reading the Components of the Genotype

The expressions of the dominance level, Expressions 2.12–2.14, depend upon the genotypic frequencies and the dominance deviation of Fisher's additive/dominance scale, d. As mentioned above, those expressions are meant to measure the effects that dominance—as defined through allele substitutions at the individual level—causes at the population level. Thus it makes sense that the expressions reflect that no (or relatively small) dominance at the population level is generated whenever there is no (or relatively small) dominance at the individual level—in fact, $d = 0$ implies that $\delta_{ij} = 0$, $ij = 11, 12, 22$ and hence also that $V_D = 0$.

From Expression 2.20, nil interaction variance (i.e., nil dominance variance, as we are not yet considering any other possible interactions) implies that all genetic variance is additive variance: $V_G = V_A$. In its turn, this implies—by definition of the broad sense heritability in Expression 2.21—that $H^2 = 1$. However, it is crucial to note that it is not justified to twist this argument conversely. Thus, we must keep in mind that (relatively) small dominance variance can occur despite significant dominance at the individual level, as shown using illustrative cases below.

We must also avoid the temptation of projecting the above argument to the additive level. Although the dominance deviations and the dominance variance depend upon the individual level only through its dominance deviation, d, the population additive effects and the additive variance depend upon not only the additive effect at the individual level, a, but also upon the dominance deviation at the individual level, d, as Expressions 2.8–2.11 show.

As mentioned above, additive parameters reflect average effects of allele substitutions, α_i, and the relative performance of individuals as breeders (under random mating) due to their genotypes, α_{ij}. It is conceptually intuitive that those properties depend upon not only the difference between the homozygotes, a, but also the presence and degree of dominance, d, as Expressions 2.8–2.10 show. Indeed, as also mentioned above, Fisher (1918) claimed to have developed such expressions to measure precisely how dominance affects those properties.

Expression 2.10 also shows how the additive effect of the gene at the population, α, depends both on a and on d. As also mentioned above, α is the slope of the linear regression to the genotypic values. Dominance can make α to be zero (i.e., the breeding values to be equal) in spite of unequal genotypic values, as shown below. The more α departs from zero, the more the breeding values will differ from each other and more effectively will selection change the genetic composition of the population. In particular, selection for high trait values makes the frequency of allele A_2 to decrease with negative values of α (as in Fig. 2.2) and vice versa.

Overall, there can be no dominance variance if there is no dominance at the individual level (i.e., $d = 0$), but similar arguments do not apply any further. Thus, for instance, the dominance variance can be zero or very low at a population in the face of high dominance at the individual level. In this regard, the additive variance can be found to work in an even more capricious way. On the one hand, similar to the dominance variance, the additive variance can be small in the face of high additive effects at the individual level. And on the other hand, beyond the way the dominance variance works, the additive variance can actually achieve high values even with nil additive effects. Although this might look somehow confusing, it indeed makes complete sense when keeping in mind that the additive and dominance variances are not designed to be constrained by simplistic logics, but instead to reflect the intricate behavior of phenotypes of complex traits in populations, as illustrated graphically though the cases discussed below.

2.3.6.6 Graphic Examples

Table 2.4 shows the values of the variables and parameters for the example used in Figs. 2.1, 2.2, 2.3, and 2.4. The values of the 10,000 phenotypes and environmental deviations are only indicated: for each genotypic class, the environmental deviations are taken from a normal distribution with zero mean and 0.1 variance: N(0,0.1). The

Table 2.4 Essential parameters and variables of the population analysis for the case in Figs. 2.1, 2.2, 2.3, and 2.4, assuming a population with $N = 10{,}000$ individuals. The environmental deviations of each class are indicated as coming from a normal distribution with a mean of zero and a variance of 0.01. Hence, each phenotype is the sum of the genotypic value and the environmental deviation of the individual. Fisher's additive dominance scale is $a = 0.5$ and $d = 0.3$, the population mean phenotype is $\mu = -1.3625$, and the additive effect of the gene at the population is $\alpha = -1$. The heritabilities are $H^2 = 0.63$ and $h^2 = 0.58$

	A_1	A_2	A_1A_1	A_1A_2	A_2A_2	Variance
Population numbers	5000	15,000	625	3750	5625	–
Population frequencies	0.25	0.75	0.0625	0.375	0.5625	–
Phenotypic level	–	–	P_I^{11}	P_I^{12}	P_I^{22}	0.2711
Genetic level	–	–	2	1.8	1	0.1711
Additive level	0.4875	−0.1625	0.975	0.325	−0.325	0.1584
Dominance level	–	–	−0.3375	0.1125	−0.0375	0.0127
Environmental level	–	–	N(0,0.1)	N(0,0.1)	N(0,0.1)	0.1

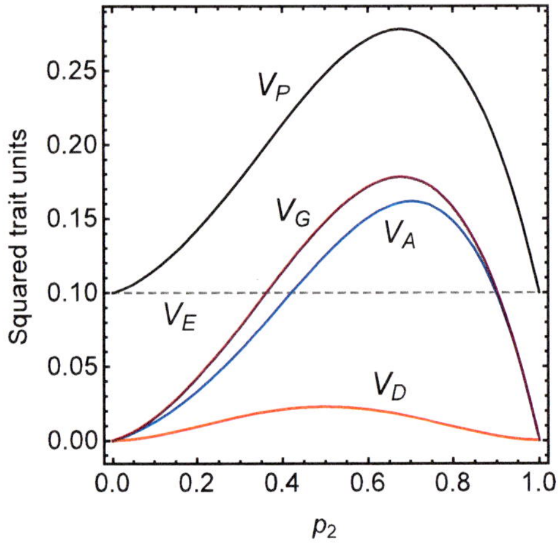

Fig. 2.5 Decomposition of the genetic variance for the one-locus two-allele case in Table 2.4, for the whole range of possible frequencies (p_2 is the frequency of A_2), assuming Hardy-Weinberg. The genetic variance (V_G, purple line) is the sum of the additive variance (V_A, blue line) and the dominance variance (V_D, red line). The phenotypic variance (V_P, black line) is the genetic variance plus the environmental variance (V_E, dashed straight line), which equals 0.1 regardless of the frequency of A_2

phenotypes are thus randomly distributed around the genotypic values, with a variance of 0.1.

The frequency of allele A_2 is $p_2 = 0.75$. Under Hardy-Weinberg proportions, the population mean phenotype is $\mu = 1.3625$ trait units. The effect of the gene at the population takes, as mentioned above, a negative value, $\alpha = -0.65$, also measured in trait units—as it can be deduced, e.g., from Expression 2.10. This value corresponds to a steep slope meaning that selection would be very effective, particularly increasing the mean phenotype by decreasing the frequency of allele A_2.

This is in accordance with a high genetic determination of the trait at the population and, specifically, of a high proportion of additive variance (see Fig. 2.5), which is measured in square trait units. Indeed, the broad and narrow sense heritabilities are $H^2 = 0.63$ and $h^2 = 0.58$, respectively (since they are ratios, they lack units). This narrow sense heritability is higher than what is commonly found in quantitative traits in practice, which must therefore have relatively higher values of variances due to interaction effects and/or environmental causes.

Indeed, the genetic basis of the example here considered is extremely simple, with one only locus and incomplete dominance (with $d = 0.3$ trait units, lower in absolute value than $a = 0.5$ trait units). We shall comment on the—as Fisher (1918) acknowledged—more realistic case of multiple loci and epistasis below. The dominance variance for the present case remains very small as compared to the additive variance not only at our population (with $p_2 = 0.75$, see Table 2.4) but for all intermediate allele frequencies, as Fig. 2.5 shows. At extreme frequencies, genetic

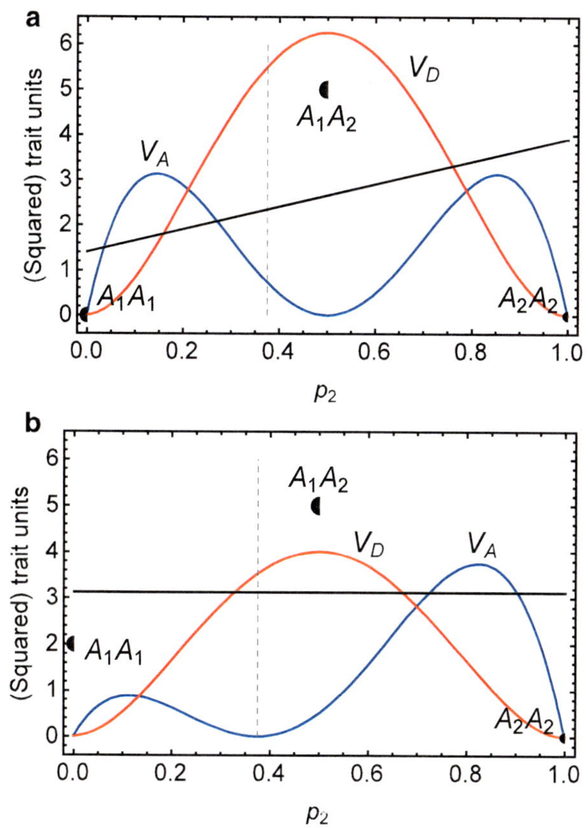

Fig. 2.6 Modified from Álvarez-Castro (2014). Decomposition of the genetic variance for a one-locus two-allele case for the whole range of possible frequencies (p_2 is the frequency of A_2), assuming Hardy-Weinberg. The additive variance (V_A, blue line) and the dominance variance (V_D, red line) are measured in square trait units. The genotypic values (black dots) are measured in trait units. The different thicknesses of the dots reflects the genotype frequencies at which the regression (black line) is performed (Hardy-Weinberg genotypic frequencies with $p_2 = 0.375$, as signaled with a grey vertical dashed line). (**a**) Nil additive effect at the individual level. The dominance effect at the individual level is $d = 5$ trait units. (**b**) Non nil additive effect at the individual level ($a = 1$ trait units). The dominance effect at the individual level is $d = 4$ trait units

variability vanishes, and, consequently, all genetic variances (V_G, V_A, and V_D) converge to zero.

Figure 2.6 shows cases with more extreme dominance interaction, particularly overdominance, which generates polymorphic equilibrium (maintenance of genetic variability) under directional selection. These cases illustrate how dominance at the individual level may either hamper or enhance selection, depending on the population frequencies. In Fig. 2.6a, with nil additive effect at the individual level ($a = 0$ trait units), all additive variance is due to the dominance effect at the individual level ($d = 5$ trait units). Other than at the fixation points ($p_2 = 0$ and $p_2 = 1$), only at

$p_2 = 0.5$, where the genetic system reaches a polymorphic equilibrium, the additive variance vanishes. The polymorphic equilibrium maintains genetic variability and thus genetic variance, all of which is dominance variance. The regression to the genotypic values shown in this figure is made at $p_2 = 0.375$, where most genetic variance is dominance variance, thus hampering selection. Nevertheless, the additive variance is not nil, enabling the system to move towards the equilibrium point. In particular, the slope of the regression is positive, $\alpha = 1.25$ trait units, indicating that selection for high phenotypes shall increase the frequency of allele A_2, p_2.

Figure 2.6b shows a similar case, with just some additive effect at the individual level this time ($a = 1$, $d = 4$ trait units). The regression, at the same point as in Fig. 2.6a, $p_2 = 0.375$, now shows a slope of zero, $\alpha = 0$ trait units, since that is precisely the polymorphic equilibrium point of this system. In this case, the maximum of additive variance is almost as high as the one of the dominance variance. The additive variance is the largest contribution to the genetic variance in a significant region to the right of the figure—with $p_2 > 0.699$. At $p_2 = 0.85$, for instance, $V_A = 3.68$ and $V_D = 1.04$, both measured in squared trait units. The dominance variance may, in fact, be much lower than the additive variance despite dominance being a crucial ingredient of the genetic system, as pointed out above.

2.3.6.7 The Resemblance Between Relatives and the Breeder's Equation

Let us at this point not lose track of the title of Fisher's (1918) foundational paper, "The correlation between relatives on the supposition of Mendelian inheritance." The correlation between relatives—i.e., between phenotypes of related individuals—is what we commonly refer to as resemblance between relatives. Now, the variance of the additive parameters at the level of the population, the additive variance, V_A, relying (as stressed right above) on both additive and dominance at the individual level (i.e., "on the supposition of Mendelian inheritance"), provides key information about the resemblance between relatives (i.e., about "[t]he correlation between relatives"). We shall hereafter, however, use the covariance instead of the correlation to reflect the resemblance between relatives since, as noted by Falconer and MacKay (1996, p 150), it is more directly related to the components of variance (the covariance of two variables equals their correlation multiplied by the product of their standard deviations).

For instance, the covariance between half sibs is (see, e.g., Falconer and MacKay, 1996):

$$cov_{HS} = \tfrac{1}{4} V_A. \tag{2.26}$$

As the previous theory, this expression has been derived assuming a simplistic context. In particular, it does neither account for a multilocus case with epistasis or departures from linkage equilibrium, nor even for a one-locus case with departures from the Hardy-Weinberg equilibrium or shared environmental effects.

Under those constraints, Expression 2.26 makes it possible to measure of V_A from empirical data on phenotypes and relatedness in a population and, hence, to obtain a measure h^2 for that population using Expression 2.25. More to the point, this kind of expressions become particularly useful in practical terms provided that they may also combine relatives of different generations. For instance, the parent/offspring covariance is (see, e.g., Falconer and MacKay, 1996):

$$cov_{\mathrm{OP}} = \tfrac{1}{2}\, V_{\mathrm{A}}. \tag{2.27}$$

Thus, it is possible to estimate V_A from phenotypes measured within the present generation using Expression 2.26 and to harness that estimate to make predictions about the phenotypes of the offspring of selected individuals (i.e., how selected individuals will perform as breeders of the next generation) using Expression 2.27. This makes it possible to ponder beforehand whether the effort required for the selection process will pay off or not.

The breeder's equation synthesizes that idea in a very simple expression:

$$R = h^2 S, \tag{2.28}$$

where R is the response to selection, measured as the difference of the mean phenotype of the offspring of the selected individuals and the original population mean phenotype, and S is the selection differential, measured as the difference between (ideally) the mean phenotype of the selected individuals and the original population mean phenotype. As noted right above, both h^2 and S can be obtained from the current population, whereas R provides a prediction of a property of the population in the next generation.

The concept of heritability and the breeder's equation were made explicit by Jay Lush (1937), and they are closely connected to the aforementioned regression towards mediocrity by Francis Galton (1886) as illustrated in Fig. 2.7. The slope of the regression line is the heritability of the trait. Since the additive variance is a portion of the total variance, the heritability is lower than one, which forces the selection response, R, to be smaller than the selection differential, S (recall Expression 2.28). The name "regression towards mediocrity" refers precisely to the fact that the resulting selection response regresses back towards the population mean phenotype relative to the selection differential applied.

The dominance variance is also meaningful in the context of resemblance between relatives, as it can be illustrated, for instance, through the covariance between full sibs (e.g., Falconer and MacKay, 1996):

$$cov_{\mathrm{FS}} = \tfrac{1}{2}\, V_{\mathrm{A}} + \tfrac{1}{4}\, V_{\mathrm{D}}. \tag{2.29}$$

This expression, in connection with either Expression 2.26 or Expression 2.27 or alternatively with Galton's (1886) regression towards mediocrity, may be used to obtain an estimation of the dominance variance from empirical data, thus disentangling the interaction component of the genetic variance, V_{D}, from the environmental variance, V_{E}. This is one of the points Fisher (1918) achieved beyond Yule's (1902) expectations, as mentioned in the previous chapter.

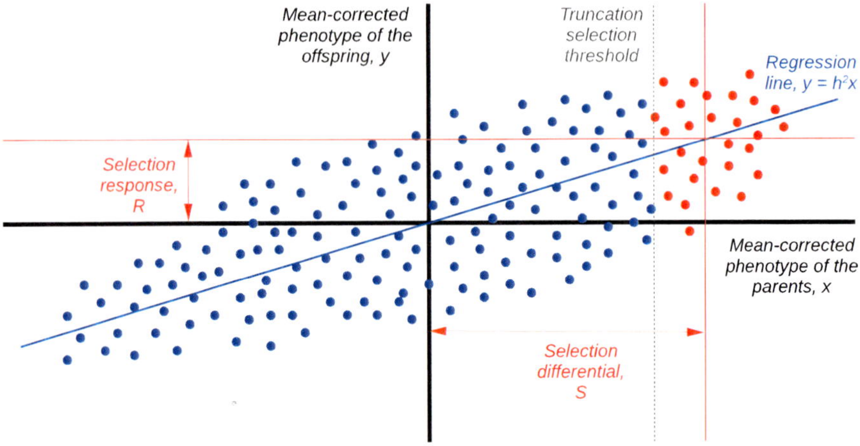

Fig. 2.7 Idealized representation of a regression towards mediocrity, in line with Falconer and Mackay's (1996) classical illustration (see their Fig. 11.1). The horizontal and vertical mains axes are centered at the mean phenotypes of parents and offspring, respectively. The horizontal and vertical red secondary axes are centered at the mean phenotype of the selected parents and their offspring, respectively (represented by the red data points). The regression line (blue line) is computed from the whole dataset (thus including blue and red data points). Its regression is lower than one, wherefore the selection response is lower than the selection differential, which justifies the label "regression towards mediocrity"

Expressions 2.26, 2.27, and 2.29 are particular cases of the general expression. Without intending to go deep into this issue, this expression is:

$$cov_{PQ} = rV_A + uV_D, \tag{2.30}$$

where $r = 2f_{PQ}$ and $u = 2f_{P1P2}f_{Q1Q2} + f_{P1Q2}f_{P2Q1}$ and where in their turn P and Q represent two individuals whose relatedness is in question; P_1, P_2, and Q_1, Q_2 represent their parents, respectively; and f stands for the co-ancestry, also called coefficient of kinship (for further details see e.g. again Falconer and MacKay, 1996).

2.4 Multiple Loci

So far, we have for simplicity dealt with one only biallelic locus under Hardy-Weinberg equilibrium. Hereafter, we shall make some comments about cases in which more loci are involved. We start by considering loci that just add up their contributions to then address epistasis.

2.4.1 Multiple Additive Loci

To begin with, just adding loci does not need to involve much higher complexity, as long as they behave in the same way as the one already considered. In what regards the influence of the multiple loci on the phenotype, we say that the loci combine their effects additively when the effect caused by allele substitutions at one locus remains constant against changes in the genetic background. Otherwise, epistasis occurs. From one locus to two additive loci, the number of parameters of Fisher's additive-dominance scale increases from three (μ, a and d) to five (μ, a^1, d^1, a^2 and d^2, where the superscripts "1" and "2" denote the two loci).

In what regards the population frequencies, we say that the loci are under linkage equilibrium when the multiple-locus frequencies are the product of the single-locus frequencies (e.g., for two loci, $p_{ijkl} = p^1_{ij} \times p^2_{kl}$). Otherwise, departures from linkage equilibrium occur, which are commonly called linkage disequilibrium. From one biallelic locus under the Hardy-Weinberg proportions to two biallelic loci under the Hardy-Weinberg proportions and linkage equilibrium, the number of parameters to describe the population frequencies increases from one, p_2, to two, p^1_2 and p^2_2.

Let us therefore consider a trait whose phenotype is the result of the action of L additive biallelic loci under Hardy-Weinberg equilibrium and linkage equilibrium. As the phenotype of each individual is the sum of the contributions of the different loci, and those contributions are statistically independent under linkage equilibrium, then the expressions of the general case can straightforwardly be obtained from the ones of the marginal loci. For instance, using superscripts for labeling the loci, with two loci A^1 and A^2, the average effect of the substitution of alleles A^1_1 and A^2_1 (i.e., of haplotype $A^1_1A^2_1$) is $\alpha_{11} = \alpha^1_1 + \alpha^2_1$ and the breeding value of individuals $A^1_1A^1_1A^2_1A^2_1$ is $\alpha_{1111} = \alpha^1_{11} + \alpha^2_{11}$. Note particularly that with two loci the average effects of allele—or, more precisely, haplotype—substitutions have two digits in the subscript, whereas the breeding values have four digits in the subscript.

In general, both the genetic parameters and variances can be obtained for the set of loci as the sum of the corresponding parameters of the marginal loci (e.g., Kempthorne, 1957). Let a trait be underlain by L loci in a population and let $h = (h_1,\ldots, h_L)$, $g = (g_1,\ldots, g_L)$ represent any haplotype and any genotype of those L loci, respectively. Then, the effect of the set of L loci, the average effect of the substitution of haplotype h, the breeding value of genotype g, and the dominance deviation of that genotype at that population can be expressed in terms of the single-locus parameters simply as, respectively:

$$\alpha_h = \sum\nolimits_{l=1}^{L} \alpha^l_{h_l}, \alpha_g = \sum\nolimits_{l=1}^{L} \alpha^l_{g_l}, \delta_g = \sum\nolimits_{l=1}^{L} \delta^l_{g_l}, \qquad (2.31)$$

And the additive, dominance, and genetic variances of that set of L loci at that population are:

$$V_A = \sum_{l=1}^{L} V_A^l,\ V_D = \sum_{l=1}^{L} V_D^l,\ V_G = \sum_{l=1}^{L} V_G^l. \tag{2.32}$$

2.4.2 Pairwise Epistasis

As noted above, epistasis occurs when the effects on the phenotype of allele substitutions made at one locus depend upon the genotypes present at other locus or loci. Fisher (1918) realized that epistasis could highly increase the complexity of his analyses. He analyzed a particular case of epistasis between two biallelic loci (a particular case of pairwise epistasis or, in his exact words, "dual epistacy") that can be described with one only parameter for epistasis (on top of the five mentioned above to describe the system without epistasis, μ, a^1, d^1, a^2 and d^2). He acknowledged that "[m]ore complex connections could doubtless exist, but the number of unknowns introduced by "dual epistacy" alone, four, is more than can be determined by existing data."

Indeed, nine parameters in total are necessary to describe a two-locus two-allele trait under (pairwise) epistasis, with nine possible genotypic values—not bound to any constraints. Fisher (1918) deemed the relatively simple case he considered with one only epistatic parameter to be illustrative of a population analysis with arbitrary epistasis. In his words: "it is very improbable that any statistical effect, of a nature other than that which we are considering, is actually produced by more complex somatic connections."

Nevertheless, it is both clear and of great interest that more complex epistatic interactions may lead to evolutionary outcomes very different from that of the simpler one considered, whether or not producing similar "statistical effects" (see, e.g., Álvarez-Castro and Le Rouzic, 2015, Fig. 8). This point is in fact particularly enlightening about Fisher's (1918) rationale, which was focused on the empirical data that could be obtained at the time. Indeed, he showed that "statistical effects" derived "on the supposition of Mendelian inheritance" explained the (empirically observed) fact of "correlation between relatives"—even though the genetic basis of the trait would remain mostly unknown.

Coming back to the four extra parameters needed to model a two-locus two-allele genetic system with arbitrary epistasis, they arise naturally from the mathematical expressions (as further explained in forthcoming chapters) and they are called additive-by-additive, *aa*; additive-by-dominance, *ad*; dominance-by-additive, *da*; and dominance-by-dominance, *dd*, epistasis (Cockerham, 1954; Kempthorne, 1954). Analogous to *a* and *d* in the one-locus case, those parameters accounting for epistasis at the individual level give rise to the corresponding parameters at the population level, $\alpha\alpha$, $\alpha\delta$, $\delta\alpha$, and $\delta\delta$. This way, the epistatic component of variance, V_I, is actually the result of four components, V_{AA}, V_{AD}, V_{DA}, and V_{DD} as (assuming the Hardy-Weinberg proportions and LE):

$$V_{\mathrm{I}} = V_{\mathrm{AA}} + V_{\mathrm{AD}} + V_{\mathrm{DA}} + V_{\mathrm{DD}}. \quad (2.33)$$

Then, Expression 2.22 for the three genotypic values of one biallelic locus turns, for the nine genotypic values of two biallelic loci with pairwise epistasis, into:

$$G_{ijkl} = \mu + \alpha^1_{ijkl} + \delta^1_{ijkl} + \alpha^2_{ijkl} + \delta^2_{ijkl} + \alpha_{\alpha ijkl} + \alpha\delta_{ijkl} + \delta\alpha_{ijkl} + \delta\delta_{ijkl}, \\ ijkl = 1111, 1112, \ldots, 2212, 2222. \quad (2.34)$$

Similarly, Expression 2.23 may with pairwise epistasis be expressed as:

$$V_{\mathrm{G}} = V^1_{\mathrm{A}} + V^1_{\mathrm{D}} + V^2_{\mathrm{A}} + V^2_{\mathrm{D}} + V_{\mathrm{AA}} + V_{\mathrm{AD}} + V_{\mathrm{DA}} + V_{\mathrm{DD}}. \quad (2.35)$$

It is critical to keep in mind that, as opposed to the multiple additive loci case, when epistasis occurs the marginal variances, V^l_{A} and V^l_{D}, $l = 1, 2$, do not coincide with the variances of the marginal cases considered independently. Nevertheless, we can still define $V_{\mathrm{A}} = V^1_{\mathrm{A}} + V^2_{\mathrm{A}}$, and $V_{\mathrm{D}} = V^1_{\mathrm{D}} + V^2_{\mathrm{D}}$. With this and Expression 2.33, Expression 2.35 takes the much simpler form:

$$V_{\mathrm{G}} = V_{\mathrm{A}} + V_{\mathrm{D}} + V_{\mathrm{I}}. \quad (2.36)$$

From Expression 2.36, Expression 2.24 with pairwise epistasis becomes just:

$$V_{\mathrm{P}} = V_{\mathrm{A}} + V_{\mathrm{D}} + V_{\mathrm{I}} + V_{\mathrm{E}}, \quad (2.37)$$

This classical nomenclature is somewhat confusing. Epistasis is gene-gene interaction and thus just one type of the possible interactions (dominance is another). However, epistasis is represented by in Expression 2.37 by an "I" for "interaction" because the "E" had previously been used for "environment." Still, we follow the classical nomenclature here because it is so conventional that it might be more confusing to try to find an alternative—which would probably still not be completely free from misinterpretation.

In any case, we may now recall that, in the one-locus case, the variance due to genetic interaction effects—i.e., the variance arising due to non-additive genetic effects—was just the dominance variance, V_{D}, as shown in Expressions 2.23 and 2.24. That remains the same in the case of multiple additive loci. Well then, Expressions 2.36 and 2.37 show that the variance arising due to non-additive genetic effects becomes with epistasis the sum of the dominance variance and the epistatic variance, $V_{\mathrm{D}} + V_{\mathrm{I}}$.

Thus, the increase in complexity when epistasis comes into play becomes already patent in the simplest two-locus two-allele case shown above, which allows for only pairwise epistasis. With more loci, the complexity of pairwise epistasis gets further increased. With three loci, for instance, the complexity increases by a factor of three—since it results in three different pairs of loci. As an example, the general additive-by-additive variance turns to be the sum of the additive-by-additive variance due to each of the three pairs of loci: $V_{\mathrm{AA}} = V^{12}_{\mathrm{AA}} + V^{13}_{\mathrm{AA}} + V^{23}_{\mathrm{AA}}$.

2.4.3 Arbitrary Epistasis and the Resemblance Between Relatives

With three loci, not only the number of possible pairwise epistatic effects gets increased, as shown right above, but also a new level of interactions may arise—third-order epistasis. The number of components of third-order epistasis is the number of different sequences of length three that can be made with the words "additive" and "dominance," i.e.: additive-by-additive-by-additive, additive-by-additive-by-dominance, additive-by-dominance-by-additive, dominance-by-additive-by-additive, additive-by-dominance-by-dominance, dominance-by-additive-by-dominance, dominance-by-dominance-by-additive, and dominance-by-dominance-by-dominance. Thus, there arise, in total, $2^3 = 8$ third-order epistatic effects.

The same rationale holds for an arbitrary number of loci. Each additional locus not only increases the number of possible interactions of the levels already present but also brings about one higher level of interactions. In general, with L loci, up to Lth-order epistasis may occur.

Thus, in mathematical terms, Mendelian inheritance with epistasis naturally brings about numerous variables. Conveniently, they also have a biological meaning, particularly in what regards Fisher's (1918) purpose of assessing the resemblance between relatives. Indeed, Expression 2.30 for the resemblance between relatives with one locus—also valid for multiple additive loci—can be expanded to three loci with arbitrary epistasis, assuming Hardy-Weinberg proportions and linkage equilibrium, in the following relatively simple manner (see, e.g., Kempthorne 1955b; Falconer and MacKay, 1996):

$$\begin{gathered} cov_{\mathrm{PQ}} = rV_{\mathrm{A}} + uV_{\mathrm{D}} + \\ r^2V_{\mathrm{AA}} + ru(V_{\mathrm{AD}} + V_{\mathrm{DA}}) + u^2V_{\mathrm{DD}} + \\ r^3V_{\mathrm{AAA}} + r^2u(V_{\mathrm{AAD}} + V_{\mathrm{ADA}} + V_{\mathrm{DAA}}) + ru^2(V_{\mathrm{ADD}} + V_{\mathrm{DAD}} + V_{\mathrm{DDA}}) + u^3V_{\mathrm{DDD}}, \end{gathered} \tag{2.38}$$

where P, Q, r, and u are as in Expression 2.30. The sequence of summands in Expression 2.38 can straightforwardly be increased to account for an arbitrary number of loci with arbitrary epistasis, with the powers of r and u at a particular summand being given by the number of times A and D, respectively, occur in the corresponding subscripts.

2.5 Fisher's Proposal in Context

All in all, the main aim Fisher (1918) pursued in his foundational paper was to enable biologists studying a heritable trait to ascertain that phenotypes and relatedness of individuals were compatible with those expected under Mendelian inheritance. It is thus noteworthy that both relatedness and phenotypes were the "observables" of the time. This of course does not mean that it is straightforward to obtain such data for any species under any conditions but just that it is at least feasible to obtain it for some species under some conditions.

Fisher's expressions provide phenotypic resemblance between relatives ultimately from the additive and dominance values of the loci involved. However, not only those values could at the time not be obtained but the loci themselves were in general unreachable, abstract entities. The simple cases for which Mendel and others after him could infer some knowledge about the causative, hereditary factors of the trait under study were precisely those simple instances that had made biologists to doubt that Mendelian inheritance would typically lead to continuous variation and gradual evolution.

2.5.1 Fisher's Breakthrough

Why then Fisher's (1918) expressions enabled such a breakthrough one century ago? On the one hand, Fisher's theory provided a decisive explanatory power, as mentioned in the previous chapter. Indeed, it provided a very broad and detailed account of how continuous traits and gradual evolution could be understood as a natural outcome of Mendelian inheritance. On the other hand, it opened doors to the development of diverse empirical studies, as further commented hereafter.

In between observable phenotypes and causative Mendelian factors, Fisher's theory steps on variance components as intermediate variables. As opposed to the high number of variables necessary for determining the genetic basis of the trait at the population, the variance components constitute a smaller set of practical statistical indicators. It thus makes sense to try to estimate variance components from data on resemblance between relatives and then use them to predict (and, thus, contrast the theory against) additional cases of resemblance between relatives. A paradigmatic example of that rationale follows.

The additive variance, V_A, is an essential indicator of phenotypic resemblance for a trait in a population that can be obtained from phenotypic data of half-sibs, using Expression 2.26. Using then Expression 2.27, the expected resemblance between one individual and one offspring can be deduced. On the whole, data from one generation makes it possible to obtain a prediction about the next generation. As pointed out above, that is particularly convenient for pondering whether a breeding program will be cost-effective or not.

The reader must have realized, though, that for the above rationale to be completely correct it is necessary not only that population conditions like Hardy-Weinberg equilibrium and linkage equilibrium are kept but also that the trait itself involves no epistasis. Indeed, Expression 2.38 shows that epistasis makes variance components other than V_{A} to be involved in the resemblances between relatives considered. Fisher (1918) disregarded problems of this kind, as mentioned above, not by disregarding epistasis to occur but by suspecting that "it is very improbable that any statistical effect, of a nature other than that which we are considering, is actually produced by more complex somatic connections."

2.5.2 Quantitative Genetics from Today's Perspective

It thus becomes crucial to keep in mind that heritability should be particularly useful under the supposition not only of Mendelian inheritance but also of Hardy-Weinberg proportions and linkage equilibrium, and even of nil or at least negligible statistical effects of second or higher order—such as V_{AA}, V_{AAA}... Falconer and MacKay (1996) had to work with two different definitions of the additive variance: "The theoretical one [...] excludes interaction variance. The practical one is that which determines, and is estimated from, covariances of relatives, and it contains fractions of the additive × additive interaction variances." Then, "...we speak of additive variance [...] referring to the practical definition with its included fractions of interaction variances."

In other words, the theory was developed based on variables that could be estimated empirically, although they might not adequately reflect their originally intended meaning. In practice, approximated predictions could be made at the cost of retaining the real meaning of the parameters sort of within a black box. Thus, a completely correct theoretical definition of the heritability was not handy because in practice the additive variance, V_A, could not be disentangled from the additive interaction variances, V_{AA}, V_{AAA}... But we must keep in mind that these interaction variances are indeed also involved in the response to selection, as it can be easily derived from Expression 2.38, which reflects that as long as r is not nil all the aforementioned variance components become involved—e.g., in the case of the covariances between parents and offspring.

In general, a radical change of perspective occurs as the underlying Mendelian factors (and, thus, computing the decomposition of the genetic variance from them) become reachable. It is indeed notable that although the genetic basis of quantitative traits was completely out of reach during the first half of the twentieth century, it is nowadays a hot topic with practical applications in agriculture, livestock, medicine, and evolutionary biology. Whether we think of quantitative trait loci, regions of the genome associated to variability of a trait of interest, marker variants to be selected in a breeding program, adaptive molecular responses, or speciation genes, variability at the level of the DNA can nowadays be scrutinized in practice. A GP map becomes thus at the least curtailed if it keeps on dealing with the genetic effects from a black box perspective alone. This fact marks a clear-cut distinction between Fisher's motivations and many of our intended use of his models today, which we simply cannot overlook when trying to upgrade and apply them. Special attention shall therefore be paid to this issue in following chapters of this book.

References

Álvarez-Castro JM (2014) Dissecting genetic effects with imprinting. Front Ecol Evol 2:51

Álvarez-Castro JM, Le Rouzic A (2015) On the partitioning of genetic variance with epistasis. In: Moore JH, Williams SM (eds) Epistasis: methods and protocols. Springer, Humana Press, New York, pp 95–114

Avery OT, Macleod CM, Mccarty M (1944) Studies on the chemical nature of the substance inducing transformation of pneumococcal types induction of transformation by a desoxyribonucleic acid fraction isolated from pneumococcus type ii. J Exp Med 79:137–158

Cockerham CC (1954) An extension of the concept of partitioning hereditary variance for analysis of covariances among relatives when epistasis is present. Genetics 39:859–882

Comstock RE, Robinson HF (1948) The components of genetic variance in populations of biparental progenies and their use in estimating the average degree of dominance. Biometrics 4:254–266

Falconer DS, Mackay TFC (1996) Introduction to quantitative genetics. Prentice Hall, Harlow

Fisher RA (1918) The correlation between relatives on the supposition of Mendelian inheritance. Trans Roy Soc Edinb 52:339–433

Galton F (1886) Regression towards mediocrity in hereditary stature. Anthropol Inst G B Irel 15: 246–263

Hershey AD, Chase M (1952) Independent functions of viral protein and nucleic acid in growth of bacteriophage. J Gen Physiol 36:39–56

Kempthorne O (1954) The correlation between relatives in a random mating population. Proc R Soc Lond B Biol Sci 143:102–113

Kempthorne O (1955a) The correlations between relatives in random matng populations. Cold Spring Harbor Symp Quant Biol 20:60

Kempthorne O (1955b) The theoretical values of correlations between relatives in random mating populations. Genetics 40:153–167

Kempthorne O (1957) An introduction to genetic statistics. Wiley, New York

Kempthorne O (1968) The correlation between relatives on the supposition of Mendelian inheritance. Am J Hum Genet 20:402–403

Lewontin RC (1974) The genetic basis of evolutionary change. Columbia University Press, Columbia

Lush JL (1937) Animal breeding plans. Iowa State College Press, Ames, IA

Moran PAP, Smith CAB (1966) Commentary on R. A. Fisher's paper on "The correlation between relatives on the supposition of Mendelian Inheritance". Cambridge University Press, Cambridge

Weinreich DM, Watson RA, Chao L (2005) Perspective: sign epistasis and genetic constraint on evolutionary trajectories. Evolution 59:1165–1174

Yule GU (1902) Mendel's laws and their probable relations to intra-racial heredity. New Phytol 1:193–207, 222–238

3 Genetic Effects Over One Century

Abstract

The study of biological inheritance led to the development of fundamental mathematical and statistical methodologies of current widespread use in general scientific practice. The achievement of the first satisfactory description of how Mendelian inheritance results in phenotypic resemblance between relatives, addressed in the previous chapter of this book, is an example of this. In this chapter, some of the most consequential advances made from the base of that theory over the last century are explored and discussed. We distinguish two types of advances. The first type aims to embrace more complex genetic architectures or more efficient ways of expressing them. The second type aims to implement more complex population conditions. The advances presented in this chapter are made in different directions and in a rather independent way. Yet, they have fertilized the ground for the development of a more ambitious framework that unifies, generalizes, and further implements them. Such framework is presented in the following chapters. Points raised in previous chapters on key warnings made about how to properly model genetic effects and on valuable lessons the history of genetics teaches us are here resumed. Keeping those points in mind will make it easier to understand the aforementioned ambitious framework in the following chapters, as it was developed in accordance with them.

3.1 Introduction

A meticulous use of the mathematical tools available was necessary for carrying out the preliminary analyses of biological inheritance, as the foundational work of Mendel bears witness to. Beyond that, a significant amount of new fundamental mathematical procedures were methodically developed for understanding and describing why and how offspring tend to resemble their parents (or, more in general, individuals tend to resemble their close relatives) more than a random

J. M. Álvarez-Castro, *Genes, Environments and Interactions*,
https://doi.org/10.1007/978-3-031-41159-5_3

individual of the population. Indeed, as stressed by Provine (1971), John B. S. Haldane (1932) successfully and more generally foresaw that:

> The permeation of biology by mathematics is only beginning, but unless the history of science is an inadequate guide, it will continue, and the investigations here summarized represent the beginning of a new branch of applied mathematics.

A paradigmatic example of how the study of inheritance fueled the development of primary mathematical tools is Galton's (1886) regression towards mediocrity, explained in the previous chapter of this book. Galton invented a method called the least squares approximation for studying the inheritance of traits of continuous variation, and, since then, it got coined as "regression" because the slope of Galton's approximation lines happened to be necessarily between zero and one, thus showing the tendency of the hereditary data to regress towards the mean. Despite the fact that many datasets may actually disperse from the mean instead (when they lead to slopes higher than one), the general method of linear least-squares keeps on being referred to as a linear regression nowadays—and the term "regression" has permeated even beyond linear approximations, generally called non-linear regressions, such as quadratic regressions.

Also in the previous chapter of this book, we have already gone through another remarkable example of a wide-ranging mathematical procedure that was originally developed to understand biological inheritance. We are referring to Fisher's (1918) analysis of variance (ANOVA), which he conceived for being able to analyze the contributions of genetic and environmental effects to the phenotypic variance, particularly through agricultural experiments at the Rothamsted the Experimental Station (now known as Rothamsed Research). Undeniably, assessing the contribution of different causative variables to the total variance of a dataset entails an invaluable tool for limitless scientific fields.

The previous list by no means ends with this and other commendable contributions by Fisher. For instance, also Karl Pearson provided in the late nineteenth and early twentieth centuries, as a foremost biometrician, numerous mathematical developments for the study of inheritance of continuous traits that are nowadays used extensively as standard statistical techniques (Provine 1971). A few examples are the development of the correlation coefficient (originally conceived by Galton), the Pearson distribution, the chi-squared test, the *p*-value, and the principal component analysis.

Yet, further significant mathematical implementations have still been required to carry on the study of biological inheritance over a century to the present, due to at least two different reasons. One of the reasons is that some mistakes were made, mainly in the interpretation of some theoretical results. This comes as no surprise, considering the extraordinary burst of knowledge generated almost in the blink of an eye. Examples of such mistakes have been pointed out in the previous chapter and several are resumed right below.

First, it was interpreted that Mendelian inheritance could not, or at least it would typically not, underlay gradual evolution of continuous traits. More in particular, Mendelian inheritance was thought not to enable selection to bring a trait beyond its observed range. Even Galton thought that his regression towards mediocrity (Galton 1886) would prevent traits of continuous variation to evolve at all. Then, Yule (1902) found out that Mendelian inheritance could actually underlay the existence and evolution of continuous variation but interpreted that interaction and environmental variances could not be told apart in practice (using empirical data on phenotypes and relatedness). Next, Fisher (1918) found out that they actually could, but interpreted that statistical effects due to epistasis, although potentially widespread, could not lead to numerically or functionally significant statistical parameters.

In its turn, Fisher's presumption about epistasis was not without risk. In fact, as already briefly pointed out in the previous chapter, those epistatic statistical effects—he regarded as negligible—condition, for instance, how the concept of heritability can be applied in practice. In perspective, we can witness to what extent Fisher's stand on epistasis originated a long-lasting (and at times intense) debate, with Sewall G. Wright pioneering, on the opposing side, the advocacy of epistasis as a crucial feature for the understanding of selection, adaptation, and evolution (Wright 1932; see also Provine 1971, 1986).

The other major reason why significant mathematical developments have been required for the study of biological inheritance along the latest century is related to the radical switch from what was observable one century ago to what can be observed nowadays. From phenotypes and relatedness, we have moved all the way to whole genome sequences and all that can be derived from them using advanced computational methods. From trying to understand and predict phenotypic correlation between relatives and the heritability of a trait, we have changed focus towards, for instance, gene mapping and the shaping agents and functional implications of complex genetic architectures. Again, it comes as no surprise that the commendable mathematical advances provided either one century ago, or half a century ago, or even quarter of a century ago (i.e., in the decade witnessing the first steps of gene-mapping experiments) needed to be revised and significantly improved. Ultimately, science progresses through a feedback between theoretical and empirical contributions.

Overall, the study of biological inheritance and particularly Fisher's (1918) theory of genetic effects laid a sound foundation for the construction of population and quantitative genetics that continued to be built upon for over one century—besides basic mathematical tools of invaluable utility for scientific research in general. Nevertheless, we still need to improve it further today, and to do so, we need to analyze in detail any remaining shortcomings. Thus, several challenges that Fisher's (1918) foundational theory of genetic effects had to face over time are reviewed in this chapter—particularly, those directly related to the theory proposed in the following chapters.

3.2 Epistasis Components

Epistasis occurs whenever the effects of allele substitutions in one locus are influenced by allele substitutions performed at other locus or loci. In terms of genetic effects, this means that epistatic effects account for the interactions that may arise between and among marginal effects—i.e., between and among the genetic effects within single loci. Since we have distinguished additive and dominance marginal effects, epistatic effects may involve all possible combinations of interactions of additive and dominance effects of different loci. This is how, for instance, a dominance-by-additive statistical effect naturally arises in a pairwise interaction or an additive-by-dominance-by-additive interaction does in a third-order epistatic interaction.

Let us focus, for instance, on the first of the two cases mentioned right above. Consider thus the (putatively nil) dominance effect, d^1 of a biallelic locus A^1, with alleles A_1^1 and A_2^1, as measured under the genetic background $A_1^2A_1^2$ of a second biallelic locus A^2. Then there is no dominance-by-additive interaction between loci A^1 and A^2—i.e., $da = 0$—as long as the dominance effect d^1 remains unchanged under the genetic background of the other homozygote of locus A^2, $A_2^2A_2^2$. Otherwise, non-nil dominance-by-additive interaction exists that makes both the dominance effect of allele of locus A^1, d^1, and the additive effect of locus A^2, a^2, to vary for certain genetic backgrounds at loci A^2 and A^1, respectively.

This has been just a taste on how the epistasis components work at the individual level. We shall revisit the effects of allele substitutions at the individual level in the next chapter. Hereafter we focus on the epistasis components at the population level, particularly through two different ways of expressing them which were published in 1954 by Oscar Kempthorne (whose work has already been useful to us in previous chapters) and his former PhD student, Clark Cockerham (having defended his PhD thesis in 1952), separately.

On the one hand, Kempthorne (1954) set a regression framework to express all the marginal and epistatic genetic effects and each "proportion of the total variance in the phenotypic values which can be attributed, in a least squares sense." to the corresponding genetic effects—i.e., the corresponding variance decomposition. His results encompass traits with multiple multiallelic loci in populations under Hardy-Weinberg equilibrium and linkage equilibrium. He particularly pointed out that "[i] t would be nice to generalize the results to the case when linkage [disequilibrium] occurs"—i.e., under departure from linkage equilibrium.

On the other hand, Cockerham (1954) provided "orthogonal scales" to obtain the genetic effects and the variance decomposition of a genetic system of two biallelic loci with arbitrary departures from Hardy-Weinberg equilibrium. He also mentioned linkage disequilibrium explicitly when pointing out that his results encompass linkage disequilibrium as long as there is no departure from Hardy-Weinberg equilibrium. Thus, Kempthorne's (1954) approach is much more flexible in what regards the genetic architecture, whereas Cockerham's (1954) widens the applicability of the study in relation to the population conditions.

Altogether, those two landmark publications highlighted the importance of epistasis for the study of correlations between relatives, thus overcoming Fisher's (1918) dismissal of this issue and improving the applicability of expressions on the correlation between relatives that were summarized in the previous chapter of this book. Kempthorne (1954) and Cockerham (1954) also coincided in noticing the importance of eventually taking linkage disequilibrium (i.e., departure from linkage equilibrium) into account. In any case, the different ranges of applicability of the two publications naturally called for a generalization (i.e., applicable to multiple multiallelic loci under arbitrary departures from Hardy-Weinerg equilibrium). Despite these works having been carried out by close collaborators, such a generalization would have to wait for more than half a century (Álvarez-Castro and Yang 2011)—and even a few more years would have to pass for the additional generalization of the theory in question with arbitrary departures from linkage equilibrium to be achieved (Álvarez-Castro and Crujeiras 2019). These generalizations are developed in detail in upcoming chapters of the present book (Chaps. 5 and 6, respectively).

3.3 Matrix Notation

In the late twentieth century, Tiwari and Elston (1997) noticed how extremely convenient it is to express the linear relationship between the genotypic values and the genetic effects using matrices and vectors—which we shall refer to as the matrix notation of models of genetic effects. In particular, they expressed Fisher's (1918) additive/dominance scale (see Table 2.1) showing that the vector of genotypic values, **G**, can be obtained by multiplying a matrix times the vector of genetic effects, **E**, as **G** = **S E**, expanding to:

$$\begin{pmatrix} G_{11} \\ G_{12} \\ G_{22} \end{pmatrix} = \begin{pmatrix} 1 & 1 & 0 \\ 1 & 0 & 1 \\ 1 & -1 & 0 \end{pmatrix} \begin{pmatrix} \mu \\ a \\ d \end{pmatrix}. \tag{3.1}$$

To be precise, Tiwari and Elston's (1997) notation choice was **F** = **A X** instead of **G** = **S E**, but we here opt to follow Zeng et al.'s (2005) nomenclature, thus referring to **S** as the genetic-effects design matrix.

In any case, for checking that Expression 3.1 actually provides the correct definitions for the additive and dominance parameters, a and d respectively, we can simply equate the vector of genetic effects using the inverse of matrix **S**, $\mathbf{S}^{-1}$, as $\mathbf{E} = \mathbf{S}^{-1}\,\mathbf{G}$, expanding to:

$$\begin{pmatrix} \mu \\ a \\ d \end{pmatrix} = \begin{pmatrix} ½ & 0 & ½ \\ ½ & 0 & -½ \\ -½ & 1 & -½ \end{pmatrix} \begin{pmatrix} G_{11} \\ G_{12} \\ G_{22} \end{pmatrix}, \tag{3.2}$$

which indeed provides $a = (G_{11} – G_{22})/2$ and $d = G_{12} – ((G_{11} + G_{22})/2)$. Additionally, $\mu = (G_{11} + G_{22})/2$ has to do with the reference from which the former two variables are measured (see Table 2.1 and Fig. 2.1).

One of the cases where the appropriateness of such matrix notation becomes especially evident is multiple loci. Indeed, a genetic-effects design matrix for two loci can be easily obtained using the single-locus one in Expression 3.1 and the Kronecker product, $\otimes$. This product operates over matrices or vectors (in general, arrays) and provides a new array whose components are obtained by multiplying the array on the right-hand side of the product times each of the scalars of the other array. For instance, if we now specify $\mathbf{S}_{1l} = \mathbf{S}$ and use $\mathbf{S}_{2l}$ to refer to the genetic-effects design matrix for two loci, then the later can be obtained from the former as $\mathbf{S}_{2l} = \mathbf{S}_{1l} \otimes \mathbf{S}_{1l}$, expanding to:

$$\mathrm{S}_{2l} = \begin{pmatrix} 1 & 1 & 0 \\ 1 & 0 & 1 \\ 1 & -1 & 0 \end{pmatrix} \otimes \begin{pmatrix} 1 & 1 & 0 \\ 1 & 0 & 1 \\ 1 & -1 & 0 \end{pmatrix}$$
$$= \begin{pmatrix} 1 & 1 & 0 & 1 & 1 & 0 & 0 & 0 & 0 \\ 1 & 0 & 1 & 1 & 0 & 1 & 0 & 0 & 0 \\ 1 & -1 & 0 & 1 & -1 & 0 & 0 & 0 & 0 \\ 1 & 1 & 0 & 0 & 0 & 0 & 1 & 1 & 0 \\ 1 & 0 & 1 & 0 & 0 & 0 & 1 & 0 & 1 \\ 1 & -1 & 0 & 0 & 0 & 0 & 1 & -1 & 0 \\ 1 & 1 & 0 & -1 & -1 & 0 & 0 & 0 & 0 \\ 1 & 0 & 1 & -1 & 0 & -1 & 0 & 0 & 0 \\ 1 & -1 & 0 & -1 & 1 & 0 & 0 & 0 & 0 \end{pmatrix}. \tag{3.3}$$

The use of the Kronecker product in this expression makes complete sense when looking at quantitative genetics from a purely mathematical point of view. In particular, the relationship between genotype and phenotype can be considered as a simple case of a tensor, as made explicit by Galas et al. (2020) from the perspective of information theory. Then the match immediately becomes clear since the Kronecker product is a particular case of a tensor product (e.g., Schafer 1996).

Now, as Tiwari and Elston (1997) also pointed out, the Kronecker product of the vectors of genetic effects gives us the key to the vector of genetic effects for two loci, as follows. Let T stand for the operation providing the transpose of a matrix or vector. We first obtain the Kronecker product of the vectors of genetic effects of the single loci,[1] $\mathbf{E}_{2l} = \mathbf{E}_2 \otimes \mathbf{E}_1 = (\mu_2, a_2, d_2)^{\mathrm{T}} \otimes (\mu_1, a_1, d_1)^{\mathrm{T}} = (\mu_1\, \mu_2, a_1\, \mu_2, d_1\, \mu_2, \mu_1\, a_2, a_1\, a_2, d_1\, a_2, \mu_1\, d_2, a_1\, d_2, d_1\, d_2)^{\mathrm{T}}$. Then we replace $\mu_1\, \mu_2$ by μ, the remaining μ_1 and μ_2 by one, and the products of single-locus additive and dominance genetic effects by the corresponding genetic effects of the epistatic interaction, leading to $(\mu, a_1, d_1, a_2, aa, da, d_2, ad, dd)^{\mathrm{T}}$. Similarly, the two-locus vector of genotypic values can be obtained as $\mathbf{G}_{2l} = (G_{1111}, G_{1211}, G_{2211}, G_{1211}, G_{1212}, G_{1222}, G_{2211}, G_{2212}, G_{2222})^{\mathrm{T}}$.

[1] Here we use an approach that is slightly different from that of Tiwari and Elston (1997), again following Zeng et al. (2005). In brief, Tiwari and Elston (1997) opted for the so-called left Kronecker product, which is obtained by multiplying the array on the left-hand side of the product times each scalar of the array on the right-hand side. We use the common Kronecker product instead, although we do the product with the arrays in the reverse order (i.e., locus two before locus A^1). This way, the genetic effects of locus A^1 appear before the ones of locus A^2 in the vector of genetic effects. Incidentally, we do not follow here Zeng et al. (2005) in the rearrangement of the columns of the resulting genetic-effects design matrices that they suggest.

Hence, for two loci the expression $\mathbf{G} = \mathbf{S}\,\mathbf{E}$—or, more precisely, $\mathbf{G}_{2l} = \mathbf{S}_{2l}\,\mathbf{E}_{2l}$—expands to:

$$\begin{pmatrix} G_{1111} \\ G_{1211} \\ G_{2211} \\ G_{1112} \\ G_{1212} \\ G_{2212} \\ G_{1122} \\ G_{1222} \\ G_{2222} \end{pmatrix} = \begin{pmatrix} 1 & 1 & 0 & 1 & 1 & 0 & 0 & 0 & 0 \\ 1 & 0 & 1 & 1 & 0 & 1 & 0 & 0 & 0 \\ 1 & -1 & 0 & 1 & -1 & 0 & 0 & 0 & 0 \\ 1 & 1 & 0 & 0 & 0 & 0 & 1 & 1 & 0 \\ 1 & 0 & 1 & 0 & 0 & 0 & 1 & 0 & 1 \\ 1 & -1 & 0 & 0 & 0 & 0 & 1 & -1 & 0 \\ 1 & 1 & 0 & -1 & -1 & 0 & 0 & 0 & 0 \\ 1 & 0 & 1 & -1 & 0 & -1 & 0 & 0 & 0 \\ 1 & -1 & 0 & -1 & 1 & 0 & 0 & 0 & 0 \end{pmatrix} \begin{pmatrix} \mu \\ a_1 \\ d_1 \\ a_2 \\ aa \\ da \\ d_2 \\ ad \\ dd \end{pmatrix}. \quad (3.4)$$

Moreover, the Kronecker product is commutable with the inverse operation (e.g, Van Loan 2000). Thus, the inverse of the genetic-effects design matrix for two loci can be obtained either by computing the Kronecker product of two single-locus inverse genetic-effects design matrices—as the one in Expression 3.2—or by inverting the two-locus genetic-effects design matrix in Expression 3.4.

And what is even more convenient, this extension of Fisher's one-locus additive-dominance scale to two epistatic loci actually entails an extension to an arbitrary number of biallelic loci with arbitrary epistasis. Indeed, the Kronecker product of design-matrices can be performed sequentially and, conveniently, the associative property holds so that, for instance, for a three-locus system it is the same to compute $(\mathbf{S}_3 \otimes \mathbf{S}_2) \otimes \mathbf{S}_1$ or $\mathbf{S}_3 \otimes (\mathbf{S}_2 \otimes \mathbf{S}_1)$—and the same holds for obtaining the vectors of the genetic effects and the vector of genotypic values following the procedure explained above. Further advantages of the matrix notation shall be presented in the following sections.

3.4 Populations in Matrix Notation

As addressed above—in the section about epistasis—models of genetic effects are required that take into account not only facts of the genetic architecture, like dominance, multiple loci, and epistasis, but also population facts, like particular allele or genotypic frequencies. As shown in the previous chapter, Fisher (1918) initially considered genotypic frequencies under the Hardy-Weinberg proportions and, particularly (in the wake of Mendel's work) those of an F_2 population, that is, with genotype frequencies $p_{11} = ¼$, $p_{12} = ½$, and $p_{22} = ¼$—for genotypes A_1A_1, A_1A_2, and A_2A_2, respectively. Zeng et al. (2005) showed that the F_2 model of genetic effects can be expressed in matrix notation as:

$$\begin{pmatrix} G_{11} \\ G_{12} \\ G_{22} \end{pmatrix} = \begin{pmatrix} 1 & 1 & -½ \\ 1 & 0 & ½ \\ 1 & -1 & -½ \end{pmatrix} \begin{pmatrix} \mu \\ a \\ d \end{pmatrix}. \quad (3.5)$$

The genetic effects can be equated from this expression as:

$$\begin{pmatrix} \mu \\ a \\ d \end{pmatrix} = \begin{pmatrix} ¼ & ½ & ¼ \\ ½ & 0 & -½ \\ -½ & 1 & -½ \end{pmatrix} \begin{pmatrix} G_{11} \\ G_{12} \\ G_{22} \end{pmatrix}. \quad (3.6)$$

As Zeng et al. (2005) explain, this is the way the genetic effects have to be defined to reflect the quantitative genetic properties of an F_2 population. We shall get into details about that in following chapters. Here we focus instead on Zeng et al.'s (2005) developments towards a more general expression, applicable in particular to arbitrary allele frequencies, p_1 and p_2—of alleles A_1 and A_2, respectively—although still restricted to genotypic frequencies under Hardy-Weinberg equilibrium. Zeng et al. (2005) call that expression the G2A model, standing for general two-allele model and expanding to:

$$\begin{pmatrix} G_{11} \\ G_{12} \\ G_{22} \end{pmatrix} = \begin{pmatrix} 1 & 2p_2 & -2p_2^2 \\ 1 & 1-2p_1 & 2p_1p_2 \\ 1 & -2p_1 & -2p_1^2 \end{pmatrix} \begin{pmatrix} \mu \\ a \\ d \end{pmatrix}. \quad (3.7)$$

Equating the genetic effects from the previous expression provides the definition of the genetic effects in the G2A model as:

$$\begin{pmatrix} \mu \\ a \\ d \end{pmatrix} = \begin{pmatrix} p_1^2 & 2p_1p_2 & p_2^2 \\ p_1 & 1-2p_1 & -p_2 \\ -½ & 1 & -½ \end{pmatrix} \begin{pmatrix} G_{11} \\ G_{12} \\ G_{22} \end{pmatrix}. \quad (3.8)$$

These single-locus expressions can be straightforwardly extended to multiple biallelic loci using the Kronecker product as explained in the previous section. Thus, the G2A model certainly broadens the scope of versatile expressions of genetic effects, while just as well evidences the convenience of even more general settings—as mentioned above, generality must be achieved in what regards both genetic architectures (e.g., multiple alleles) and population properties (like departures from Hardy-Weinberg equilibrium and from linkage equilibrium). Nevertheless, before addressing such challenge (in following chapters), we shall comment on yet another advantage of the matrix notation—the change-of-reference operation.

3.5 Change of Reference

Already during the mid twentieth century, Van der Veen (1959) compared different ways to model the genetic effects, which he referred to as different "metrics." In the context of the models he worked with, he perceived Fisher's additive/dominance scale (Expressions 3.1 and 3.2) as the simplest way of expressing the genetic effects. Nevertheless, in the end, all models are "different ways of describing the genotypic values." He pointed out that each way reflects different properties. Thus, the same way as the F_2 model (Expressions 3.5 and 3.6) reflects the quantitative properties of an F_2 population, Fisher's additive/dominance scale reflects the quantitative properties of a population in which half of the individuals are A_1A_1 and the other half are A_2A_2—i.e., with genotype frequencies $p_{11} = ½$, $p_{12} = 0$, and $p_{22} = ½$. Those proportions are the ones resulting from infinite generations of selfing over an F_2 population and that for this reason it makes sense to refer to Fisher's additive/dominance scale as the F_∞ model (Van der Veen 1959).

In any case, Van der Veen (1959) provided a table (Table 1 in his publication) "[f]or translation of genotypic values (or functions of genotypic values) from one metric into another," i.e., for translating the genetic effects resulting from a model into the ones resulting from another model. The different models use different starting points (i.e., references) to measure the genetic effects—that reference point is μ in Expressions 3.1, 3.2, and 3.4–3.8. This is why when this concept was later on taken up (or rediscovered) and developed further, it was called change-of-reference operation (Hansen and Wagner 2001b).

Now, the matrix notation makes the change of reference straightforward, as shown hereafter. Let $\mathbf{S}_M$ and $\mathbf{E}_M$ be the genetic-effects design matrix and the vector of genetic effects of model M, respectively. Then, expressions of the genetic effects of Fisher's additive/dominance scale (i.e., the F_∞ model) and those of the F_2 model (particularly, Expressions 3.1 and 3.3, respectively) become, respectively: $\mathbf{G} = \mathbf{S}_{\mathrm{F2}}\,\mathbf{E}_{\mathrm{F2}}$, $\mathbf{G} = \mathbf{S}_{\mathrm{F}\infty}\,\mathbf{E}_{\mathrm{F}\infty}$. Both are just expressions of the same genetic effects, $\mathbf{G}$, in terms of different genetic effects. Then, by equating $\mathbf{G}$ in those equalities, it follows $\mathbf{S}_{\mathrm{F2}}\,\mathbf{E}_{\mathrm{F2}} = \mathbf{S}_{\mathrm{F}\infty}\,\mathbf{E}_{\mathrm{F}\infty}$. Finally, by equating $\mathbf{E}_{\mathrm{F2}}$ in this resulting equality, it is straightforward to obtain the genetic effects of the F_2 model in terms of those of the F_∞ model as $\mathbf{E}_{\mathrm{F2}} = \mathbf{S}_{\mathrm{F2}}^{-1}\,\mathbf{S}_{\mathrm{F}\infty}\,\mathbf{E}_{\mathrm{F}\infty}$.

For a given genetic architecture (a set of genotypic values), the genetic effects of the F_∞ model reflect properties of an F_∞ population. We have just shown how to obtain from them the corresponding properties of the same genetic architecture in an F_2 population. This can be generalized to other populations. If we have models of genetic effects "1" and "2" expressed in matrix notation, the general expression of change of reference from "1" to "2" is (Álvarez-Castro and Carlborg 2007):

$$\mathbf{E}_2 = \mathbf{S}_2^{-1}\,\mathbf{S}_1\,\mathbf{E}_1. \tag{3.9}$$

This expression is broad in itself since it can potentially be applied to any two populations. It must be kept in mind, however, that Expression 3.9 cannot be applied unless the genetic-effects design matrices of the populations in question are available. Thus, it becomes evident once again that it is necessary to develop models of genetic effects that account for the most general situations possible—both in what regards the genetic architectures and the population structures.

3.6 Individual and Population Models

As mentioned in the former section, Fisher's additive/dominance scale, although originally designed as a simple way to reflect properties of a genetic system at the individual level, also provides quantitative genetic parameters of certain populations—particularly, those called F_∞ populations with genotypic frequencies

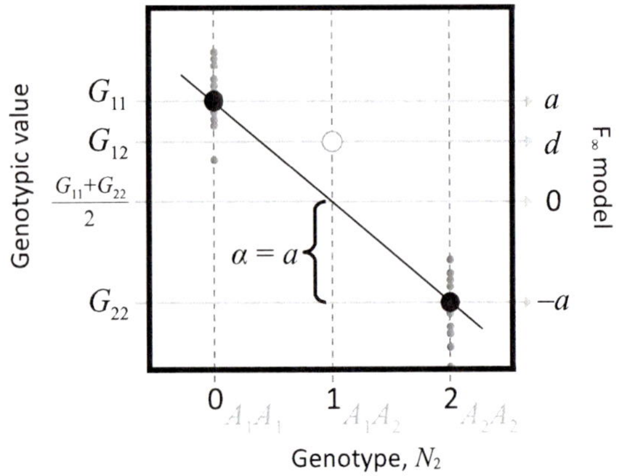

Fig. 3.1 Regression of the raw phenotypic values of the individuals to the gene content (of allele A_2, N_2) in an F_∞ population, naturally resulting in the interpolation to the genotypic values of the homozygotes (solid black line). The slope of that line is the additive effect of the gene at an F_∞ population. The genotypic value of the heterozygote, G_{12}, is shown void to reflect that $p_{12} = 0$

$p_{11} = ½$, $p_{12} = 0$, and $p_{22} = ½$. But then, this becomes fairly disturbing since Fisher (1918) precisely stressed how detrimental it would be to "obscure the essential distinction between the individual and the population," as already discussed in the previous chapter of this book. Let us thus devote the present section to analyze that issue in detail.

3.6.1 The Case of the F_∞ Model

In the previous chapter, we have illustrated that the additive effect of a gene at a population is the slope of the linear regression of the genotypic values on the gene content, i.e., on the number of A_2 alleles of each individual (see Fig. 2.2). Now, in an F_∞ population, with no heterozygotes ($p_{12} = 0$), no such regression is possible. Nevertheless, an interpolation can be performed instead, providing the only line passing through the genotypic values of the two homozygotes. Incidentally, that line can also be obtained as the regression of the raw phenotypic values to the gene content (see Fig. 3.1).

Note indeed that when the regression of the genotypic values can be performed (i.e., with $p_{ij} \neq 0$, $ij = 11, 12, 22$), it necessarily results in the same regression line as the regression of the raw phenotypic values to the gene content. Hence, the latter actually provides the additive effect of the gene at the F_∞ population, α, as shown in Fig. 3.1. And thus, such population-wise additive effect coincides with the individual-wise additive effect of Fisher's additive/dominance scale (see Fig. 2.1).

Note also that, in an F_∞ population, the lack of heterozygotes makes it impossible to obtain a measure of the dominance value, d. Therefore, for the change of reference from an F_∞ to an F_2 population exemplified above ($\mathbf{E}_{F2} = \mathbf{S}_{F2}^{-1}\, \mathbf{S}_{F\infty}\, \mathbf{E}_{F\infty}$) to make complete sense, information about the genetic architecture that cannot be inferred from the F_∞ population itself would have to be obtained. In any case, we here highlight that it is noteworthy that despite Fisher's (1918) deep concern about making a clear-cut distinction between the individual and the population, his additive/dominance scale can be interpreted in a population context. Indeed, in spite of Fisher's additive/dominance scale having been designed to reflect properties at the level of the individual, it is nowadays commonly called the F_∞ model, thus stressing precisely the population properties reflected by its parameters.

3.6.2 Other Ambivalent Models

Thus, we have shown how Fisher's additive/dominance scale is related to two different meanings. With this in mind, it will be easier to understand that there are other models that fall under this same condition. In particular, these are the ones with the same regression slope as Fisher's additive/dominance scale. Mathematically expressed, such models are the ones with either $p_{11} = p_{22}$ or $p_{12} = 0$ (as derived by Álvarez-Castro and Carlborg 2007). The F_∞ population fulfills both conditions. The F_2 population does not fulfill the second one, but it fulfills the first, which means that it can also be potentially understood both as a population model and as an individual model, in a similar way as Fisher's additive/dominance scale (i.e., the F_∞ model).

In the context of comparing genetic architectures obtained in mapping experiments (potentially using different models in the analyses), Cheverud and Routman (1995) also designed a model focusing on the "physiological" dimension rather than biased towards any population reference. Their proposal was to assign no differential population frequencies to the genotypes. Nevertheless, assuming no differential genotype frequencies can also be understood as fitting to a population with equal genotype frequencies, $p_{11} = p_{12} = p_{22} = 1/3$ (which fulfills the first condition above, $p_{11} = p_{22}$). Thus, again, Cheverud and Routman's (1995) model can be understood both as a population and as an individual model, as Fisher's additive/dominance scale. In the second sense, it leads to an unweighted regression of the genotypic values on the gene content. Indeed, the model went eventually called the unweighted regression (UWR) model by Zeng et al. (2005), who expressed it in matrix notation as:

$$\begin{pmatrix} G_{11} \\ G_{12} \\ G_{22} \end{pmatrix} = \begin{pmatrix} 1 & 1 & -1/3 \\ 1 & 0 & 2/3 \\ 1 & -1 & -1/3 \end{pmatrix} \begin{pmatrix} \mu \\ a \\ d \end{pmatrix}. \tag{3.10}$$

To sum up, some of the previous genetic models (specifically, both the F_∞, the F_2, and the UWR models) may be understood to have an interpretation at the level of the individual because the slope of their regression of the genotypic values on the gene content is $\alpha = a$, while all of them (including now the G2A model as well) provide genetic effects measured from a reference, μ, that can be understood as the phenotype mean of a population. Thus, some of the models analyzed above could have some interpretation at the level of the individual, but even all of those also have an interpretation at the population level. Therefore, for the clear-cut distinction between the individual and the population Fisher (1918) stressed as crucial to be made, an additional approach would have to be developed.

3.6.3 Models of Genetic Effects Exclusively at the Individual Level

Hansen and Wagner (2001b) proposed to develop genetic effects reflecting allele substitutions performed from individual genotypes, thus fitting to a clear and useful biological interpretation at the individual level. As mentioned above, they actually brought Van der Ween's (1959) transformation concept to the level of a change-of-reference operation, providing a method to change the reference from which the genetic effects are measured from one individual genotype to any other, and also to the mean of a population—thus providing genetic effects as average effects of allele substitutions in the context of a population.

Hansen and Wagner's (2001b) proposal of individual-wise expressions of genetic effects was innovative in what regards its neat biological interpretation. Indeed, it has a more direct interpretation of genetic effects at the level of the individual (since the genetic effects are effects of allele substitutions from the reference of individual genotypes). Besides, its expressions do not entail any duplicate interpretation of the kind of Fisher's additive/dominance scale (i.e., the F_∞ model), the F_2 model, and the UWR model. More specifically, Hansen and Wagner's (2001b) proposal cannot be understood as reflecting by itself properties of any given population, as explained right below.

If a model of genetic effects using the reference of an individual genotype was intended to reflect the properties of a population, such population would necessarily consist only of individuals of that one only genotype, e.g., with $p_{11} = 1$, $p_{12} = p_{22} = 0$ (assuming A_1A_1 as the reference individual genotype). But then it would not be possible to perform any regression (or interpolation) in that population since we need at least two points to determine a line. The mathematics and the genetics perspectives are coherent since it is also true that no genetic effects make full sense at such population—at which any differences in phenotype would be due to environmental effects alone. Thus, Hansen and Wagner's (2001b) finally proposed a way of expressing the genetic effects directly and purely focused on the individual level. Since their developments were initially restricted to multilinear epistasis, additional developments would be required to achieve a higher level of generality, which will be addressed in the following chapters.

3.7 The Quest for a General Theory of Genetic Effects

Genetics stands on the shoulders of the gigantic contribution of the early study of biological inheritance (which comprised the development of essential mathematic methodologies). In turn, numerous commendable contributions were made to bring Fisher's (1918) seminal models of genetic effects to higher ranges of theoretical elucidation and practical applicability. Several of them are presented in this chapter. However, several others are still required. It is necessary to fully integrate the theory of genetic effects into the shared core of the evolutionary and quantitative genetics corpuses, particularly in order to enable a more adequate use of the astonishing possibilities of phenotypic and genotypic data made accessible in recent times. Such additional implementations will be addressed in the following chapters.

3.7.1 The First Implementations

The many implementations of the conditions initially considered by Fisher (1918) can be classified into two major groups. On the one hand, an effort was made in properly analyzing complex genetic architectures including both larger (with more alleles and loci), more complex (with more and more efficiently modeled genetic and gene-environment interactions) genetic architectures, and more efficient ways of mathematically expressing them (particularly, using the matrix notation). On the other hand, further developments had also to be provided for broadening the scope of the properties of the populations considered, particularly aiming for models fitting non-equilibrium situations like departures from the Hardy-Weinberg equilibrium and from linkage equilibrium, and gene-environment correlations.

As explained above, neither one group nor the other (and thus let alone both simultaneously) were always systematically addressed. Instead, many implementations were made following independent lines of progress. This way of proceeding was paradigmatically exemplified by close collaborators Kempthorne (1954) and Cockerham (1954) having developed models of genetic effects with epistasis, entailing multiple multiallelic loci at equilibrium populations and two biallelic loci enabling departure from Hardy-Weinberg equilibrium, respectively.

3.7.2 Matrix Notation and Change of Reference

Tiwari and Elston's (1997) matrix notation provided a basis for more systematic generalizations, entailing an extension from single to multiple loci. Zeng et al. (2005) further generalized the use of the matrix notation by building up genetic-effects design matrices that implement arbitrary allele frequencies. Indeed, their G2A model does not only account for a particular genetic architecture in a particular population (with particular population frequencies) but for multiple biallelic epistatic loci in populations under the Hardy-Weinberg equilibrium and linkage equilibrium. In the following chapters, we shall further generalize this setting to account for, e.g., multiple alleles, gene-environment interactions, and arbitrary departures from equilibrium frequencies.

Models of genetic effects must be generalized in what regards population frequencies because we need genetic effects that reflect properties at the population level, which therefore are different for the same genetic architecture in different populations. It then becomes convenient to be able to derive the properties of a genetic architecture in a population from those in a different population. This procedure is called change of reference, and it was initially developed at the individual level—for obtaining the genetic effects as effects of allele substitutions from one individual genotype given those from another individual (Hansen and Wagner 2001b).

Above, we have shown that Tiwari and Elston's (1997) matrix notation also enables a convenient way to perform the change of reference operation—as long as genetic-effects design matrices have been derived for the genetic architecture in question in the populations implied in the aimed change of reference. The name of the change of reference operation makes it explicit that the genetic effects are always attached to a reference point, which can be an individual or a population average. The change of reference has been analyzed from different perspectives (e.g. Hansen and Wagner 2001a; Barton and Turelli 2004; Álvarez-Castro and Carlborg 2007; Mani et al. 2008; Pavlicev et al. 2010; Álvarez-Castro and Yang 2011; Álvarez-Castro et al. 2012; Le Rouzic et al. 2013; Álvarez-Castro and Le Rouzic 2015). In any case, it must be kept in mind that neither the matrix notation nor the change-of-reference operation provides *per se* generalizations of the models of genetic effects. As mentioned above, such generalizations are addressed in the next chapters of this book.

3.7.3 Revisiting Fisher's Remarks

Our quest for a general theory of genetic effects must keep in mind the concerns raised by Fisher (1918) in relation with his foundational proposal (which have been discussed in the previous chapter). The first remark was about the polarity Fisher (1918) noticed his models implicitly entailed. Specifically, he perceived that the models were somehow anticipating that one particular allele was associated to higher phenotypic values than the other. All models dealt with in this chapter suffer from that same preconception. We shall provide a solution for this preconception in the next chapter.

The second remark was about Fisher's (1918) awareness of his developments to stem from allele substitutions. In particular, he meant his additive/dominance scale to provide a yardstick reflecting allele substitutions performed over genotypes—thus, at the individual level. He apparently failed to foresee the inconvenient consequences coming from the fact that such allele substitutions actually fit squarely those performed in the context of a population. Indeed, it is paradoxical—even ironical—that his intended population-independent yardstick eventually turned to be commonly known under the name of a population—the F_∞ population. As explained above, this issue may be overcome by developing truly population-independent yardstick precisely describing the effects of allele substitutions performed over specific individual genotypes, as proposed by Hansen and Wagner (2001b). We shall deal with such models in the context of general genetic architectures in the next chapter.

As presented above, Fisher's second remark overlaps with the third one—avoiding confusion between the individual and the population. Since his intended allele substitutions at the individual level can also (even more appropriately) be interpreted at the population level, Fisher fell into the trap he was forewarning others of. Also in the next chapter, we shall address this issue. Indeed, by providing developments for a general way to model allele substitutions from individual genotypes, we shall also establish a clear theoretical distinction between individual and population parameters and, in its turn, a solid ground for a broader scope of biological interpretations.

3.7.4 Overcoming Hurdles of the Past

To end up with, we here recall the background considerations we found in Chap. 1 to be also worth keeping in mind when addressing our current challenge on providing a general and satisfactory theory of genetic effects. The reader may want to keep them in mind particularly throughout the next three chapters. Nevertheless, they will be resumed in perspective in Chap. 9.

Our first and very general consideration was about new proposals, whether critical with established courses of action, deserving objective scrutiny. Second and much more specifically, genetic interactions, in spite of significantly increasing the complexity of a study, have shown decisive explanatory power since the very beginnings of the study of biological inheritance, thus becoming hazardous to carelessly disregard them. Third and back to general terms, the pace of scientific enhancement significantly slows down whenever personality clashes are not left aside. And fourth, particularly in the face of the previous difficulties, it is crucial to keep an eye on any possible favorable contexts in order to take advantage of them for disentangling even the tightest conceptual knots.

In this chapter, we have exposed and discussed some of the most consequential advances made along one century on models of genetic effects. We have focused, in particular, on those that most watered the ground to make flourish a unifying and generalizing framework. Such a framework is presented mainly in the next three chapters. Throughout both those chapters and the remaining ones, the convenience and in fact the necessity of that framework will be argued. One of its advantages is to enable a more meticulous examination of Fisher's (1918) suspicion about statistical effects of gene interactions—a subject that has raised significant scientific disagreement during the last decades (see e.g., Hansen, 2013; this issue is resumed in Chaps. 6 and 9). This chapter has fulfilled its purpose if it helps the reader to realize that the theory presented in the following chapters is worth developing.

References

Álvarez-Castro JM, Carlborg Ö (2007) A unified model for functional and statistical epistasis and its application in quantitative trait loci analysis. Genetics 176:1151–1167

Álvarez-Castro JM, Crujeiras RM (2019) Orthogonal decomposition of the genetic variance for epistatic traits under linkage disequilibrium-applications to the analysis of Bateson-Dobzhansky-Muller incompatibilities and sign epistasis. Front Genet 10:54
Álvarez-Castro JM, Le Rouzic A (2015) On the partitioning of genetic variance with epistasis. In: Moore JH, Williams SM (eds) Epistasis: methods and protocols. Springer, Humana Press, New York, pp 95–114
Álvarez-Castro JM, Yang R-C (2011) Multiallelic models of genetic effects and variance decomposition in non-equilibrium populations. Genetica 139:1119–1134
Álvarez-Castro JM, Carlborg O, Ronnegard L (2012) Estimation and interpretation of genetic effects with epistasis using the NOIA model. Methods Mol Biol 871:191–204
Barton NH, Turelli M (2004) Effects of genetic drift on variance components under a general model of epistasis. Evol Int J Org Evol 58:2111–2132
Cheverud JM, Routman EJ (1995) Epistasis and its contribution to genetic variance components. Genetics 139:1455–1461
Cockerham CC (1954) An extension of the concept of partitioning hereditary variance for analysis of covariances among relatives when epistasis is present. Genetics 39:859–882
Fisher RA (1918) The correlation between relatives on the supposition of Mendelian inheritance. Trans Roy Soc Edinburgh 52:339–433
Galas DJ, Kunert-Graf J, Uechi L, Sakhanenko NA (2020) Towards an information theory of quantitative genetics. bioRxiv 811950
Galton F (1886) Regression towards mediocrity in hereditary stature. Anthrop Inst Great Britain and Ireland 15:246–263
Haldane JBS (1932) The causes of evolution. Longmans, Green and Co, London
Hansen TF (2013) Why epistasis is important for selection and adaptation. Evolution 67:3501–3511
Hansen TF, Wagner GP (2001a) Epistasis and the mutation load. Genetics 158:477–485
Hansen TF, Wagner GP (2001b) Modeling genetic architecture: a multilinear theory of gene interaction. Theor Popul Biol 59:61–86
Kempthorne O (1954) The correlation between relatives in a random mating population. Proc R Soc Lond B Biol Sci 143:102–113
Le Rouzic A, Álvarez-Castro JM, Hansen TF (2013) The evolution of canalization and evolvability in stable and fluctuating environments. Evol Biol 40:317–340
Mani R, St Onge RP, Hartman JLT, Giaever G, Roth FP (2008) Defining genetic interaction. Proc Natl Acad Sci U S A 105:3461–3466
Pavlicev M, Le Rouzic A, Cheverud JM, Wagner GP, Hansen TF (2010) Directionality of epistasis in a murine intercross population. Genetics 185:1489–1505
Provine WB (1971) The origins of theoretical population genetics. University of Chicago Press, Chicago
Provine WB (1986) Sewall Wright and evolutionary biology. University of Chicago Press, Chicago
Schafer RD (1996) An introduction to nonassociative algebras. Dover, New York
Tiwari HK, Elston RC (1997) Deriving components of genetic variance for multilocus models. Genet Epidemiol 14:1131–1136
Van Der Veen JH (1959) Tests of non-allelic interaction and linkage for quantitative characters in generations derived from two diploid pure lines. Genetica 30:201–232
Van Loan C (2000) The ubiquitous Kronecker products. J Comput Appl Math 123:85–100
Wright S (1932) The roles of mutation, inbreeding, crossbreeding and selection in evolution. In 6th International congress of genetics, pp 356–366
Yule GU (1902) Mendel's laws and their probable relations to intra-racial heredity. New Phytol 1:193–207, 222–238
Zeng ZB, Wang T, Zou W (2005) Modeling quantitative trait Loci and interpretation of models. Genetics 169:1711–1725

4 Genetic Architectures at the Individual Level

Abstract

Mendel explained, in the second half of the nineteenth century, the inheritance of certain traits as the action of pairs of factors—one-locus two-allele genetic systems—with complete dominance. Since then, many other schemes underlying the inheritance of traits—i.e., other genetic architectures—have been considered and discovered. Consequently, increasingly complex mathematical models of genetic effects needed to be developed. In this chapter, we present models fitting a potentially infinite range of genetic architectures. We here restrict our scope to models whose parameters reflect phenotypic effects of allele substitutions at the individual level—i.e., as effects of allele substitutions from the reference of an individual genotype. Particularly, we address one biallelic locus, multiple alleles, multiple loci, gene-environment interaction, sex-linked loci, gene-sex interaction, and imprinting. Some of the models here developed have not been published previously. Indications on how to proceed when several of the previous situations occur simultaneously (e.g., multiple alleles and multiple loci) are provided. The models here presented enable us to better tackle the distinction between the individual and the population levels as well as several other fundamental issues raised at the birth of quantitative genetics. Besides, they will serve as a basis to address the models at the population level in the next chapters.

4.1 Introduction

Establishing a mathematical description of the connection between genotype and phenotype (allowing for the influence of the environment) was necessary to merge Mendel's theory of inheritance and Darwin's theory of evolution over one century ago, as discussed in the previous chapters. The resulting synthetic theory of evolution became a jewel of scientific thought. Its explanatory power soon enabled applications to real data analysis. In the mid-twentieth century, the discovery of

J. M. Álvarez-Castro, *Genes, Environments and Interactions*,
https://doi.org/10.1007/978-3-031-41159-5_4

the double helix of DNA (Franklin and Gosling 1953; Watson and Crick 1953) opened the door to an exponential growth of molecular genetics applications. About four decades later, gene mapping experiments became feasible. It then became evident that the original genotype-to-phenotype (GP) map needed to be upgraded.

Today, it may look paradoxical that the GP map was for a long time developed as a black box in what regards the effects of alleles and genotypes on phenotypes—the applications of the models remained significantly blind not only to the actual effects of the alleles within the genes and their possible interactions but even to the number of genes implied. Nevertheless, such a curtailed strategy entailed a necessary virtue since it provided a practical way to bridge the not-yet-observable genetic effects. Indeed, it enabled predictions of practical use based on data that, at least for some populations, was obtainable at that moment—phenotypes and relatedness. As explained in the previous chapters, the applications dealt with estimates of variances of genetic and environmental effects rather than with those effects themselves, since the variances (but not the number of loci and their effects), could be estimated from relatedness and phenotypes.

As early as the 1930s, Fisher found the red cell blood groups to directly reflect information of the molecular level (Crouch and Bodmer 2020). Nevertheless, the first preliminary step towards mapping loci underlying a quantitative trait in a model species would come have to wait for the arrival of the 1960s (Thoday 1961) and the technical advances to obtain the data necessary for estimating the genetic effects themselves eventually developed mostly from the 1990s. It then became evident, as mentioned above, the need to adapt the mathematical models to locate, estimate, and explicitly use genetic effects. Thus, it became necessary to develop expressions for the variables to be estimated to meet the requirements of the statistical estimation procedures, which can occur in two different ways.

First, we want no constraints in our variables that could prevent us from detecting potential facts of the real genetic architectures, like certain kinds of gene interactions. For instance, a model that does not consider imprinting can obviously not provide estimates of imprinting. And second, we want our models to be able to fit the genotypic frequencies of the data since any mismatches between the way the variables are entered in the models and the genotype frequencies of the data would result in a biased estimation of the genetic effects. For instance, a model that does not consider departures from the Hardy-Weinberg proportions will potentially provide biased estimates whenever the data used are not under the Hardy-Weinberg proportions.

On the other hand, the use of the models of genetic effects for locating, estimating, and explicitly using genetic effects also reinforced the need to precisely determine the biological meaning of the variables representing the genetic effects in the models. Such improvement is indeed key to a sound biological interpretation of the estimates obtained from the data. Tuning the model to the data is, as has just been mentioned, the proper statistical way to obtain the initial estimates, which have a biological meaning precisely in the context of the data used. The models have to be flexible enough to provide, from those initial estimates, the ones with the desired biological meaning. The desired estimates are the ones fitting the question(s) we would like to answer in our study—e.g., how large is the additive effect of a particular allele substitution made on a particular genetic/environmental background. The procedure of transforming a set of genetic effects fitting a biological meaning into a new set fitting a different biological meaning is known as a change-of-reference operation. It has been shown in the previous chapter and it shall be further considered in the present one.

For all the cases mentioned, new mathematical developments were required. Although those developments could not be of significant practical use in the context of the applications carried out before gene-mapping experiments were feasible, they became crucial from that point on. Several important developments made to adapt the GP map to gene-mapping experiments were discussed in the previous chapter. Namely, the way epistasis had been implemented in the original theory since the 1950s (Cockerham 1954; Kempthorne 1954) was expressed using a matrix notation (Tiwari and Elston 1997); variables accounting for population frequencies were included in such notation framework (Zeng et al. 2005) and the concepts of individual-based genetic effects and change-of-reference between sets of genetic effects reflecting different biological meanings were established (Hansen and Wagner 2001).

In this and the following chapters, we will build on those advances towards a general GP map, i.e., we will gradually provide and discuss the necessary implementations to upgrade the GP map in the aforementioned senses. To begin with, in the present chapter, we provide models of genetic effects at the individual level (individual-referenced models) and add up on the change-of-reference operation. The understanding of these models will help us resume and further dissect several remarks already raised by Fisher (1918) when he first disclosed his theory of genetic effects—as the need of a clear-cut distinction between the individual and the population levels. On the whole, we will hereafter address crucial issues about the meaning of the parameters of the GP map under different genetic architectures, focusing in particular on the individual level.

4.2 One Biallelic Locus

We shall in what follows adhere to Hansen and Wagner's (Hansen and Wagner 2001) logic and describe the genetic effects as effects of allele substitutions made from one individual genotype, thus clarifying the biological meaning of genetic effects at the individual level. The matrix notation—commented in the previous chapters—shall prove consistent with that aim and make important operations much easier. With a view to a completely general formulation, we shall first consider the simplest case of a phenotype underlain by one biallelic locus, A, with alleles A_1 and A_2—the situation considered by Álvarez-Castro and Carlborg (2007)—and then gradually broaden the scope towards more complex genetic architectures.

Let us take genotype A_1A_1 as a starting point. Thus, the reference, R, to start measuring the effects of allele substitutions is the genotypic value of A_1A_1, which can be expressed as:

$$R = G_{11}. \tag{4.1}$$

From that reference, the effect of substituting allele A_1 by allele A_2 can be obtained by comparing the expected phenotype of A_1A_1 (i.e., the genotypic value G_{11}) against the expected phenotype of A_2A_2 (i.e., the genotypic value G_{22}). Specifically, the effect of substituting allele A_1 by allele A_2 is:

$$a = \frac{G_{22} - G_{11}}{2}, \tag{4.2}$$

since it is necessary to make two substitutions of allele A_1 by allele A_2 in order to get from A_1A_1 to A_2A_2. The name "additive" and the notation a make reference to the fact that each allele substitution is assumed to add up the same effect to the phenotype.

We have hitherto deliberately avoided dealing with the heterozygote, A_1A_2, and with its corresponding genotypic effect, G_{12}. We could actually have felt tempted to define the effect of one allele substitution just as $G_{12} - G_{11}$, which would avoid dividing by two as in Expression (4.2). However, it is well known that this would not be convenient since the expected phenotype of the heterozygote, G_{12}, not only differs from those of the homozygotes, G_{11} and G_{22}, not only due to the mere effect of having made an allele substitution, but also due to the interaction that may emerge between two different alleles that become present at the same time only in the heterozygote.

There is, therefore, a second effect to consider—the interaction that may emerge by two different alleles being present in the same genotype (that of the heterozygote), as compared against the absence of interaction. Such comparison may at first glance look tricky to measure, since we do not have a means to have two alleles together in the same genotype while preventing them from interacting. Conveniently, however, we can from Expressions 4.1 and 4.2 infer how the two alleles would behave without interaction. Indeed, assuming no interaction effect, the genotypic value of heterozygote A_1A_2 would be just $R + a = G_{11} + a$, which using Expression 4.2 can be written as $G_{11} + (G_{22} - G_{11})/2 = (G_{11} + G_{22})/2$. Thus, the interaction effect is the departure of G_{12} from this value, that is:

$$d = G_{12} - \frac{G_{11} + G_{22}}{2}. \tag{4.3}$$

The name dominance and the notation d have been chosen following Mendel.

Expressions 4.1–4.3 provide the reference point and the additive and dominance effects, respectively. There are now two ways in which those expressions can be reformulated using matrix notation. Let T stand for the transpose operation, $\mathbf{G} = (G_{11}, G_{12}, G_{22})^{\mathrm{T}}$ be the vector of genotypic values, and $\mathbf{E} = (R, a, d)^{\mathrm{T}}$ be the vector of genetic effects, which also include the reference point, R. Then, on the one hand, we can use Expressions 4.1–4.3 to express the vector of genetic effects in terms of the vector of genotypic values under the form $\mathbf{E} = \mathbf{S}^{-1}\,\mathbf{G}$ (a notation introduced in the previous chapter), as:

Table 4.1 Obtaining the genetic effects for pea color (Mendel's trait), at the individual level (adapted from Álvarez-Castro 2016). Mendel's hereditary factors (alleles) are A_1 and A_2. The genotypic value of genotype A_iA_j, G_{ij}, is its expected phenotype. The phenotypes are seed color measured as "yellowness" (thus, zero for the green peas and one for the yellow ones). The genotype of the green peas, A_1A_1, is chosen as reference. Thus its genotypic value, G_{11}, becomes the reference point, R. From there on, the table shows how the genetic effects are derived. Figure 4.1 provides a graphical interpretation of this process.

Genotype	Seed color	G_{ij}	Genetic effects (including reference point)
A_1A_1	Green	0	$R = G_{11} \Rightarrow R = 0$
A_2A_2	Yellow	1	$R + 2a = G_{22} = 1 \Rightarrow a = (1 - R)/2 \Rightarrow a = ½$
A_1A_2	Yellow	1	$R + a + d = G_{12} = 1 \Rightarrow d = 1 - R - a \Rightarrow d = ½$

$$\begin{pmatrix} R \\ a \\ d \end{pmatrix} = \begin{pmatrix} 1 & 0 & 0 \\ -½ & 0 & ½ \\ -½ & 1 & -½ \end{pmatrix} \begin{pmatrix} G_{11} \\ G_{12} \\ G_{22} \end{pmatrix}. \tag{4.4}$$

It is reassuring that we can obtain this expression also by describing the genotypic values in terms of the reference point and the additive and dominance effects, as follows. Expression 4.1 provides the reference point, R, as the genotypic value of the homozygote A_1A_1 (G_{11}) and, thus, conversely, also the way round, that is, $G_{11} = R$. Then, the genotypic value of the heterozygote, A_1A_2, (G_{12}) can be obtained as G_{11} plus one allele substitution (i.e., an additive effect, a) plus (since the two alleles come to be present at the same time) a dominance interaction effect, d, that is, $G_{11} = R + a + d$. Finally, the genotypic value of the remaining homozygote, A_2A_2, (G_{22}) can be obtained as G_{11} plus two additive effects, $2a$, that is, $G_{11} = R + 2a$. On the whole, we can write the following expression of the form $\mathbf{G} = \mathbf{S}\ \mathbf{E}$:

$$\begin{pmatrix} G_{11} \\ G_{12} \\ G_{22} \end{pmatrix} = \begin{pmatrix} 1 & 0 & 0 \\ 1 & 1 & 1 \\ 1 & 2 & 0 \end{pmatrix} \begin{pmatrix} R \\ a \\ d \end{pmatrix}. \tag{4.5}$$

It is now easy to verify that we have just obtained an expression equivalent to Expression 4.4. Indeed, the matrix, $\mathbf{S}^{-1}$, in Expression 4.5 actually is the inverse of the matrix, $\mathbf{S}$, in Expression 4.4. In other words, Expression 4.5 can be obtained by just equating the vector of genotypic values from Expression 4.4—and, analogously, Expression 4.4 can be obtained by equating the vector of genetic effects from Expression 4.5. Table 4.1 and Fig. 4.1 illustrate how the genetic effects at the individual level are obtained using the well-known example of color seed Mendel studied in peas (following Álvarez-Castro 2016).

4.3 Being Consistent with Fisher's Remarks

In the previous chapters, we highlighted three remarks Fisher made in his seminal publication (Fisher 1918). The first one is about a certain polarity of the genetic effects as he originally defined them. The second one is about recallig that these

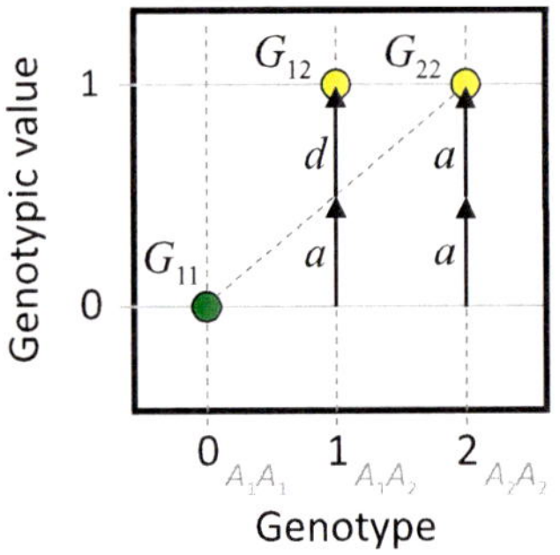

Fig. 4.1 Graphical representation of the process explained in Table 4.1 (adapted from Álvarez-Castro 2016). The genotypes are in the horizontal axis, which measures the number of A_2 alleles. The genotypic values, G_{ij}, are represented in the vertical axis. The arrows indicate how the genetic effects are measured

genetic effects reflect effects of allele substitutions. The third one is the need to keep in mind a clear-cut distinction between the individual and the population levels. Hereafter, we show that the above developments Expressions 4.1–4.5—and, thus, also the forthcoming ones based on them—are entirely consistent with these remarks, in fact considerably more so than Fisher's (1918) own developments.

4.3.1 Polarity

Let us refer to the model developed above as F_{G11} (since it uses G_{11} as a reference point) and to Fisher's additive/dominance scale as F_∞ (for the reasons discussed in the second half of the previous chapter). Fisher (1918) noticed that the way he defined the additive effect in the F_∞ implied a polarity. As Table 4.1 shows, Fisher's (1918) choice for the additive effect at the individual level, a, is half the distance from the homozygote for the "second" allele, A_2, to the "first" allele, A_1, $(G_{11} - G_{22})/2$. In brief, Fisher (1918) defined a as the "negative" distance between the "first" homozygote and the "second" one. Assuming the polarity that the A_1 performs better (it has a higher phenotype) than A_2 is in line with that definition, which makes a to measure the absolute value of an assumed decrease.

Fisher (1918) was aware that overdominance could already distort such assumed performance (as the performance of the heterozygote would not be on the way from one homozygote to the other one). Besides, he acknowledged that epistasis could be widespread, in which case the assumed polarity could get reversed under a different genetic background, as already pointed out in Chap. 2. Ultimately, interactions turn the polarity inherent in Fisher's developments uncomfortable (to say the least). But this must not have been a great concern for Fisher after all, as he disregarded the importance of interactions in the action of selection, as also pointed out in the previous chapters (and it will be resumed in the following ones).

The choice for a in the F_{G11} above (Expression 4.2) is just minus the one in Fisher's (1918) F_∞ (see Table 4.1). With just this choice, the F_{G11} is free from any imposed polarity—the additive effect, a, is not presupposed to be either positive, or negative, or nil. Instead, it is naturally positive when going from the "first" homozygote, A_1A_1, to the "second" one, A_2A_2, increases the phenotype; it is negative when it decreases it; and it is nil when it does not change it. Overall, no by-default assumption is made about the relative performance of the alleles in Expressions 4.1–4.5 and, thus, they do not suffer from structural limitations for the study of interactions.

4.3.2 Allele Substitutions

The polarity addressed under the previous heading involved the comparison between the performance (the phenotype) of the different genotypes (with particular focus on the homozygotes). Thus, we have implicitly discussed the polarity Fisher noticed in his models in terms of the effects of allele substitutions. As mentioned above, Fisher also made it explicit that the genetic effects necessarily reflect effects of allele substitutions. Our above developments (Expressions 4.1–4.5) not only reflect effects of allele substitutions but they are actually developed as explicit effects of allele substitutions. In particular, the parameters a and d are defined as unambiguous effects of allele substitutions from particular genotypes, i.e., at the individual level. They are in particular not interpretable as averages over populations, which actually brings us to the third and last point.

4.3.3 The Distinction Between the Individual and the Population Levels

We have already addressed the distinction between the individual and the population levels in some detail in the previous chapters. In what follows, we shall further elaborate on this distinction in the light of the individual-referenced model developed above. Please recall first of all that, the genetic effects, a and d, in Expressions 4.2 and 4.3) are defined almost in the same way as in the F_∞ model (cf. Fig. 2.1 and related text in Chap. 2). The only difference is the change in sign of a, as discussed right above—by just slightly modifying the F_∞ (by changing the sign in the definition of a) both F_{G11} and F_∞ would share the same genetic effects (see Table 4.2).

However, F_{G11} and F_∞ are different models beyond that sign difference. Even if correcting it, the biological meaning of their genetic effects is different. In what regards the biological interpretation, this difference goes so far as the former is an individual-referenced model, whereas the latter fits a population context—it can be understood as a population-referenced model. Indeed, in F_{G11} the genetic effects are explicitly defined as the effects of allele substitutions from the reference of a particular genotype, as explained above, whereas in F_∞ they reflect average effects

Table 4.2 Genetic effects (including the reference point) of the F_{G11} model Expressions 4.1–4.5 and the F_{∞} model (Fisher's (1918) additive dominance scale, Fig. 2.1, Table 2.1 and related text in Chap. 2), in terms of the genotypic values, as they can be deduced from the expression $\mathbf{G} = \mathbf{S}\,\mathbf{E}$ (Expression 4.5 above for F_{G11}, and Expression 3.2 for F_{∞})

	Reference point	Additive effect, a	Dominance effect, d
F_{G11}	G_{11}	$\frac{G_{22}-G_{11}}{2}$	$G_{12}-\frac{G_{11}+G_{22}}{2}$
F_{∞}	$\frac{G_{11}+G_{22}}{2}$	$\frac{G_{11}-G_{22}}{2}$	$G_{12}-\frac{G_{11}+G_{22}}{2}$

Table 4.3 Decomposition of the F_{G11} and the F_{∞} models, in terms of the genotypic values, as they can be deduced from the expression $\mathbf{E} = \mathbf{S}^{-1}\,\mathbf{G}$ (Expression 4.4 above for F_{G11} and Expression 3.1 for F_{∞})

	G_{11}	G_{12}	G_{22}
F_{G11}	R	$R + a + d$	$R + 2a$
F_{∞}	$\mu + a$	$\mu + d$	$\mu - a$

of allele substitutions in the context of a particular population, as introduced in Chaps. 2 and 3 and further commented below.

The F_{G11} and the F_{∞} models are not only supposed to be interpreted in different biological ways, they are in fact very different mathematically as well. Indeed, the two models have different reference points—F_{G11} uses a particular genotype as reference whereas F_{∞} uses the midpoint between the two homozygotes, which is the mean of an F_{∞} population (see Table 4.2). Having different reference points makes the two models to actually behave in very different ways—they lead to completely different decompositions of the genotypic values, as shown right below. This holds even if we make the additive and dominance genetic effects to coincide with just the aforementioned sign change.

Table 4.3 shows that, despite the similarities shown in Table 4.2, the decomposition of the genotypic values of the F_{G11} and the F_{∞} models are indeed very different. If we modify the sign of a in F_{∞} for its genetic effects to completely coincide with those of F_{G11}, as commented above, then the sign of a in the last row of Table 4.3 would just get modified accordingly (i.e., the decompositions of the genotypic values of the homozygotes would simply get interchanged). Thus, regardless of the sign difference in the genetic effects, the different reference points of the two models make them to lead to completely different decompositions of the genotypic values, which in turn provide different insights into the biological properties of the trait.

In what regards the F_{G11} model, the decompositions explain the genotypic values as a set of allele substitutions from the reference genotype ($R = G_{11}$), as developed above. Concerning now the F_{∞} model, the interpretation of the decomposition of the genotypic values makes real sense at the population level. As explained in Chap. 3, it was Van Der Veen (1959) who pointed out that the decomposition of the genotypic values resulting from Fisher's additive/dominance scale provided directly the population properties for a population with genotype frequencies $p_{11} = ½$, $p_{12} = 0$, and $p_{22} = ½$. With Mendel's experiments in mind, such frequencies may be thought of as the result of a large number (ideally, infinity) generations of selfing from an F_2 population—hence the label F_∞.

Thus, when taken as a whole, Fisher's (1918) additive/dominance scale matches a population-based interpretation rather than an individual-based one. This is remarkable (and actually paradoxical) given that Fisher originally developed his additive/dominance scale as a yardstick, thus regardless any population reference. We have already explained in Chap. 2 that Fisher's additive/dominance scale was developed at a time when the genetic effects themselves where somewhat abstract entities and designed for just expressing the genotypic values in the simplest way possible that makes the dominance interaction explicit. Fisher even meant the genotypic values to be reparameterized so that $\mu = 0$ (see Table 2.1, Fig. 2.1 and related text in Chap. 2), i.e., to get to simply $G_{11}^{ad} = a$, $G_{12}^{ad} = d$, and $G_{22}^{ad} = -a$.

If we now replace a and d by α and δ, respectively, the decomposition of the genotypic values of the F_∞ model (see Table 4.3) takes the form $G_{11} = \mu + \alpha$, $G_{12} = \mu + \delta$, and $G_{22} = \mu - \alpha$. This decomposition can be obtained in a population context from Expressions 2.8–2.10, 2.13, and 2.22, just using the aforementioned genotypic frequencies $p_{11} = ½$, $p_{12} = 0$, and $p_{22} = ½$. The corresponding decomposition of the genetic variance (Expression 2.23) can also be derived using the theory developed in Chap. 2 (Expressions 2.11 and 2.14).

It is worth noting that the decompositions of F_{G11} model could be derived analogously, with genotypic "frequencies" $p_{11} = 1$, $p_{12} = 0$, and $p_{22} = 0$. However, two comments must be made. First, and expectedly, the decomposition of the genotypic values so obtained shows a difference in the sing of the additive effect as compared to the one shown in Table 4.3, for the reasons already explained above. Second, and reassuringly, all genetic variances so obtained are nil regardless of the values the genetic effects may take—i.e., the F_{G11} model is not, in fact, a population model.

Ultimately, there are two key-points why Fisher's additive/dominance scale fits an F_∞ population. First, the reference point chosen by Fisher in order to get to the simplest expressions possible actually is the mean of an F_∞ population, $\mu = (G_{11} + G_{22})/2$. Second, in an F_∞ population, the average effects of allele substitutions happen to coincide with substitutions made at the individual level (the level Fisher had in mind when developing his additive/dominance scale). As explained in the previous chapter, this coincidence occurs whenever either $p_{11} = p_{22}$ or $p_{12} = 0$ (Álvarez-Castro and Carlborg 2007). In an F_∞ population, not just one but actually both conditions are met.

Beyond Fisher's additive/dominance scale (i.e., the F_∞ model), the aforementioned conditions are met, for instance, by Cheverud and Routman's (1995) physiological model (fitting a population with $p_{11} = p_{12} = p_{22} = 1/3$) and by the F_2 model (see Chap. 3). All of them have been proposed as common grounds to express and compare results of mapping experiments. However, none of them uses an individual genotype as reference. Thus, all of them reflect the biological properties of the genetic system at a certain population determined by a certain set of genotypic frequencies.

Therefore, for dealing with the individual level, we adhere to purely individual-referenced models, whose parameters unambiguously reflect allele substitutions made from individual genotypes (Hansen and Wagner 2001; Álvarez-Castro and Carlborg 2007). As for the models whose reference points do not coincide with the individual-referenced models, we will treat all of them as population-referenced models for all intends and purposes, regardless of whether their genetic effects coincide with the ones of the individual-referenced models or not. The notation used for the two types of models will reflect this difference in order to avoid confusion. Thus, we shall from here on use Latin letters (such as R, a, and d) for reference points and genetic effects of the individual-referenced models and Greek letters (such as μ, α, and δ) for all population-referenced models, including those (like the F_∞ model) whose genetic effects coincide with the ones of the individual-referenced models.

4.4 Change of Reference

The procedure followed above for developing and individual-referenced model using G_{11} as reference point—the F_{G11} model—can certainly be followed to obtain also models using G_{12} and G_{22} as reference points, i.e., the F_{G12} and the F_{G22} models. By doing so, we would obtain the genetic-effects design matrices of these two models as:

$$\mathbf{S}_{G12} = \begin{pmatrix} 1 & -1 & -1 \\ 1 & 0 & 0 \\ 1 & 1 & -1 \end{pmatrix}, \mathbf{S}_{G22} = \begin{pmatrix} 1 & -2 & 0 \\ 1 & -1 & 1 \\ 1 & 0 & 0 \end{pmatrix} \tag{4.6}$$

and their corresponding inverse matrices as:

$$\mathbf{S}_{G12}^{-1} = \begin{pmatrix} 0 & 1 & 0 \\ -½ & 0 & ½ \\ -½ & 1 & -½ \end{pmatrix}, \mathbf{S}_{G22}^{-1} = \begin{pmatrix} 0 & 0 & 1 \\ -½ & 0 & ½ \\ -½ & 1 & -½ \end{pmatrix}. \tag{4.7}$$

Comparing Expressions 4.4, 4.7, it follows that the three models, F_{G11}, F_{G12}, and F_{G22}, lead to the same expressions for the additive and dominance genetic effects, a and d (since the second and three rows of the three matrices are identical). Thus, the change-of-reference operation as introduced in the previous chapter (Expression 3.9) is somehow superfluous in what regards transforming vectors of genetic effects within individual-referenced models—they would basically transform the reference point to the new model (the model with the new reference) and leave the additive and dominance effects unchanged. The change-of-reference operation is thus much more useful to transform individual-referenced genetic effects into population-referenced genetic effects, and vice versa, and also to transform population-referenced genetic effects between different populations.

In any case, for the individual-referenced models, it is possible to transform genetic-effects design matrices as well (Álvarez-Castro and Carlborg 2007), as:

$$\mathbf{S}_2 = \mathbf{S}_1 - \mathbf{R}_2\,\mathbf{S}_1\,\mathbf{I}_0, \tag{4.8}$$

where $\mathbf{S}_1$ is the known genetic-effects design matrix (say, $\mathbf{S}_{G11}$), $\mathbf{S}_2$ is the genetic-effects design matrix we want to obtain (say, $\mathbf{S}_{G12}$), $\mathbf{R}_2$ is the matrix whose columns indicate the new reference point (for obtaining $\mathbf{S}_{G12}$, it would be a 3×3 matrix filled with zeros but for the second column, filled with ones), and $\mathbf{I}_0$ is an identity matrix, but for its first element (at its upper-left corner), which is zero.

With Expression 4.8, it is possible to obtain a general expression of the genetic effects design matrix, i.e., an expression that provides the genetic-effects design matrix for any reference point, as a function of indicator variables. For the case we have so far considered (one biallelic locus), such function can be expressed as [1]:

$$\mathbf{S} = \begin{pmatrix} 1 & -b_{12} - 2b_{22} & -b_{12} \\ 1 & 1 - b_{12} - 2b_{22} & 1 - b_{12} \\ 1 & 2 - b_{12} - 2b_{22} & -b_{12} \end{pmatrix}, \tag{4.9}$$

where b_{11}, b_{12} and b_{22}, are indicator (binary) variables (e.g., $b_{11} = 0$, $b_{12} = 1$, and $b_{22} = 0$, for obtaining $\mathbf{S}_{\mathrm{F}G12}$).

4.5 Multiple Alleles

Multiallelic models of genetic effects (whether individual-referenced of population-referenced) entail a significant increase in complexity in relation to biallelic models. Indeed, already the number of genetic effects is to be understood in detail in order to make a correct use of these models (see, e.g., Yang and Álvarez-Castro 2008). In the biallelic models, one only additive genetic effect (a or α, for the individual- and population-referenced models, respectively) is accounted for the effects of allele substitutions between alleles A_1 and A_2. We shall in this section consider an arbitrary number of alleles, r. Let us start by considering the simplest case of a multiallelic model, i.e., with $r = 3$.

We have thus alleles A_1, A_2, and A_3 and, therefore, three possible pairs of different alleles, A_1A_2, A_1A_3, and A_2A_3. Correspondingly, there are three (potentially) different additive effects, which we can represent (in the context of an individual-referenced model) as a_{12}, a_{13}, and a_{23}, respectively. Now, these three parameters account, respectively, for substitutions between alleles A_1 and A_2, A_1 and A_3, and A_2 and A_3. But precisely because the effects of allele substitutions are additive (i.e., because the additive genetic effects are actually additive), a_{12} and a_{13} necessarily provide a_{23}

[1] Álvarez-Castro and Carlborg (2007) used p_{ij} to denote the indicator variables—b_{ij} in Expression 9—in the context of comparing the individual-referenced models with the population-referenced models and, for instance, derive the aforementioned conditions for the additive and dominance effects of the population-referenced models to coincide with those of the individual-referenced models ($p_{11} = p_{22}$, or $p_{12} = 0$). We have here changed the notation for the sake of avoiding any possible confusion between individual-referenced and population-referenced models, which is one major goal of this book.

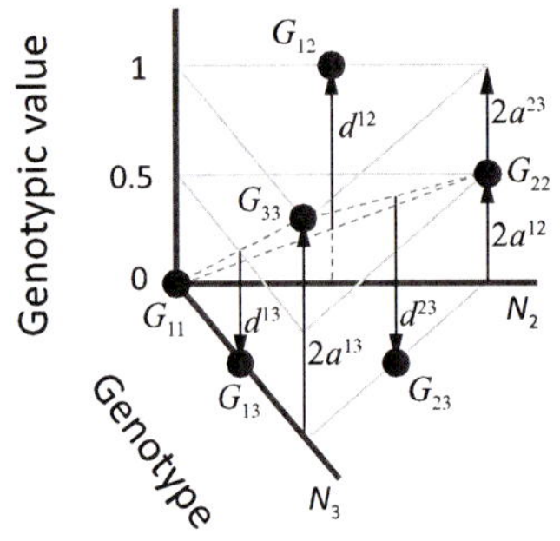

Fig. 4.2 Graphical representation of the individual-referenced genetic effects of a one-locus three allele model. The genotypic values (G_{ij}) are in the vertical axis and the genotypes are in the horizontal axes, as in Fig. 4.1, with two horizontal axes in this case, accounting for the numbers of A_2 and of A_3 alleles (N_2 and N_3 axes, respectively). As in the two-locus case, the difference between the genotypic values of the two homozygotes determines the additive effect (of substituting allele A_1 by allele A_2) and the dominance effect is the departure of the heterozygote from its additive expectation (due to the presence of interaction between the two alleles), for each pair of alleles. We have chosen a case in which allele A_1 displays complete dominance over allele A_3, alleles A_1 and A_2 display overdominance, and alleles A_2 and A_3 display underdominance

as $a_{23} = a_{13} - a_{12}$. Note though that this reasoning does not apply to the dominance effects since each pair of alleles could (or not) display a different kind of interaction (e.g., complete dominance, underdominance, no dominance . . .).

Overall, with three alleles, we would have five independent genetic effects, a_{12}, a_{13}, d_{12}, d_{13}, and d_{23}, which, accounting also for the reference point, R, build up a vector of genetic effects, **E**, with six scalars. This is as expected, since it equals the number of genotypic values, G_{11}, G_{12}, G_{22}, G_{13}, G_{21}, and G_{33} (see Fig. 4.2). This way, it will keep on being possible to connect the vector of genotypic values, **G**, with the vector of genetic effects by means of a 6×6 square genetic-effects design matrix, as $\mathbf{G} = \mathbf{S}\ \mathbf{E}$. Let us thus obtain such matrix.

We shall actually start by obtaining the inverse of the genetic-effects design matrix, $\mathbf{S}^{-1}$. Indeed, from Expressions 4.4, 4.7, it is already possible to infer how to build $\mathbf{S}^{-1}$ for three alleles. The first row just indicates the reference point of choice. Regardless of the reference point chosen, the remaining rows follow the definition of the genetic effects in terms of the genotypic values. Such definitions are as in Expressions 4.2 and 4.3, i.e.:

$$aij = \frac{G_{jj} - G_{ii}}{2}, dij = G_{ij} - \frac{G_{ii} + G_{jj}}{2}, i,j = 1,2,3, i < j. \tag{4.10}$$

Then the order of the genetic effects in **E** provides the order of the rows and the order of the genotypic values in **G** provides the column in which each scalar of each row has to be placed. Thus, assuming, for instance, $R = G_{11}$, the expression $\mathbf{E}_{G11} = \mathbf{S}_{G11}^{-1}\ \mathbf{G}$ becomes:

$$\begin{pmatrix} R \\ a^{12} \\ d^{12} \\ a^{13} \\ d^{13} \\ d^{23} \end{pmatrix} = \begin{pmatrix} 1 & 0 & 0 & 0 & 0 & 0 \\ -½ & 0 & ½ & 0 & 0 & 0 \\ -½ & 1 & -½ & 0 & 0 & 0 \\ -½ & 0 & 0 & 0 & 0 & ½ \\ -½ & 0 & 0 & 1 & 0 & -½ \\ 0 & 0 & -½ & 0 & 1 & -½ \end{pmatrix} \begin{pmatrix} G_{11} \\ G_{12} \\ G_{22} \\ G_{13} \\ G_{23} \\ G_{33} \end{pmatrix}. \quad (4.11)$$

We have used superscripts instead of subscripts in order to prevent confusion between the genetic effects in this expression and the partition of the genotypic values (see, e.g., Expression 2.22). Preventing such confusion of course makes more sense in the context of the models at the population level which we shall address in the next chapters—we already use the same superscripts here for consistency.

Now by just inverting $\mathbf{S}_{G11}^{-1}$ in Expression 4.11, we get to the equivalent expression $\mathbf{G} = \mathbf{S}_{G11}\mathbf{E}_{G11}$, expanding to:

$$\begin{pmatrix} G_{11} \\ G_{12} \\ G_{22} \\ G_{13} \\ G_{23} \\ G_{33} \end{pmatrix} = \begin{pmatrix} 1 & 0 & 0 & 0 & 0 & 0 \\ 1 & 1 & 1 & 0 & 0 & 0 \\ 1 & 2 & 0 & 0 & 0 & 0 \\ 1 & 0 & 0 & 1 & 1 & 0 \\ 1 & 1 & 0 & 1 & 0 & 1 \\ 1 & 0 & 0 & 2 & 0 & 0 \end{pmatrix} \begin{pmatrix} R \\ a^{12} \\ d^{12} \\ a^{13} \\ d^{13} \\ d^{23} \end{pmatrix}. \quad (4.12)$$

The genetic-effects design matrix $\mathbf{S}_{G11}$ can also be obtained directly, in the same way as in the biallelic case above (Expression 4.5 and related text).

All the above has a straightforward extension to an arbitrary number of alleles, r. Of the total number of $r(r-1)/2$ additive effects, the independent $r-1$ entering the model are a^{1i}, $i = 2, \ldots, r$. The remaining additive effects can be obtained from them as $a^{ij} = a^{1j} - a^{1i}$. The definition of the genetic effects and the genetic-effects design matrix and its inverse follow the very same logic as in Expressions 4.10–4.12. Expression 4.8 for obtaining genetic-effects design matrices also applies to the multiallelic case. As in the biallelic case, Expression 4.8 also enables, for any given number of alleles, a general genetic-effects design matrix (expressed in terms of indicator variables of the genotypes), thus analogous to Expression 4.9. Indeed, such matrix for the three-allele case was provided by Yang and Álvarez-Castro (2008) as:

$$\mathbf{S} = \begin{pmatrix} 1 & -2b_2 & -b_{12} & -2b_3 & -b_{13} & -b_{23} \\ 1 & 1-2b_2 & 1-b_{12} & -2b_3 & -b_{13} & -b_{23} \\ 1 & 2-2b_2 & -b_{12} & -2b_3 & -b_{13} & -b_{23} \\ 1 & -2b_2 & -b_{12} & 1-2b_3 & 1-b_{13} & -b_{23} \\ 1 & 1-2b_2 & -b_{12} & 1-2b_3 & -b_{13} & 1-b_{23} \\ 1 & -2b_2 & -b_{12} & 2-2b_3 & -b_{13} & -b_{23} \end{pmatrix}, \quad (4.13)$$

where b_{13} and b_{23} are the indicator variables of genotypes A_1A_3 and A_2A_3, respectively, $b_2 = ½b_{12} + b_{22} + ½b_{13}$ and $x_3 = ½b_{13} + ½b_{23} + b_{33}$.

By comparing Expressions 4.5 and 4.12, it can be seen that the genetic-effects design matrix for r alleles is an extension of the one for $r - 1$ alleles—the biallelic matrix in Expression 4.5 is at the upper-left corner of the matrix in Expression 4.12. By looking at those expressions (and even to Expression 4.13) it is easy to infer how that extension can be done recursively up to any number of alleles (i.e., how to build the four remaining square submatrices, the easiest of which is the one in the upper-right corner, completely filled with zeros). Such procedure may be of great computational use and has been described in detail by Yang and Álvarez-Castro (2008).

4.6 Multiple Loci

The way to build individual-referenced models with multiple loci has already been addressed in the previous chapter. Two one-locus models expressed in matrix notation can be combined into a two-locus model using an array operation called Kronecker product. The Kronecker product of the vectors of genotypic values of the two loci provides the order of the two-locus genotypic values in the two-locus vector of genotypic values, which also applies to the vectors of genetic effects. The Kronecker product of the genetic-effects design matrices of the two loci directly provides the two-locus genetic-effects design matrix, which can also be done with the inverses of the genetic-effects design matrices.

The construction of models of multiple loci by means of the Kronecker product applied to single-locus models has many advantages. For instance, loci can be added to the model sequentially, meaning, for instance, that a three-locus model can be built by applying the indications above to a two-locus model and a one-locus model. A four-locus model can either be built from a three-locus model and a one-locus model or from two two-locus models. In any case, variables accounting for all possible genetic interactions between/among loci—the epistasis components—naturally arise in the multiple-locus resulting models, as also illustrated in the previous chapter.

The aforementioned interactions make the dimension of the multiallelic models to increase significantly. In a one-locus two-allele model (e.g., Expression 4.5), we have three genotypic values, three genetic effects (including the reference point), and

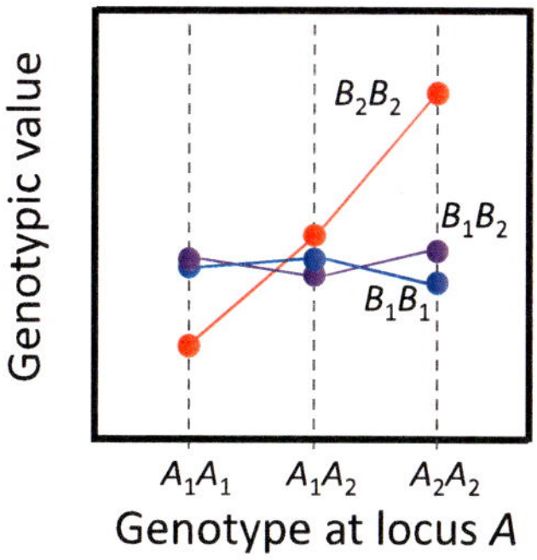

Fig. 4.3 Two-locus two-allele GP map with epistasis. As the lines are not parallel, epistasis occurs. In this case, locus A is canalized (the effects of allele substitutions are small) by allele B_1 at locus B. In general, epistasis occurs as long as the lines are not parallel (see a similar example in Álvarez-Castro 2016)

a 3×3 genetic-effects square matrix. Without within-locus interactions (i.e., dominance), the dimension would shrink to two, with the genotypic value of the heterozygote being obtainable from those of the homozygotes ($G_{12} = (G_{11} + G_{22})/2$) and a nil dominance interaction ($d = 0$). In a general two-locus two-allele model (e.g., Expression 3.4, although mind that this is not a purely individual-referenced model in the sense discussed above), the dimension rises to $3^2 = 9$. Without between-locus interactions (i.e., without epistasis), the dimension would shrink to five, as four genotypic values would be obtainable from the other four (e.g., $G_{2222} = G_{1111} + (G_{1122} + G_{2211})/2$) and the four epistasis genetic effects being equal to zero ($aa = da = ad = dd = 0$). The dimension of a general three-locus two-allele model rises to $3^3 = 27$.

As shown above for multiple alleles (Fig. 4.2), the complexity of the multiple-locus models becomes evident also when depicting them graphically. Still, Cheverud and Routman (1995) used a relatively simple (only two-dimensional) way for visualizing the nine genotypic values of a two-locus two-allele model (see Fig. 4.3). That graphical representation provides a way to grasp the pattern of interaction, even if the values of the genetic effects reflecting the interactions are more difficult to infer visually.

When initially introducing the application of the Kronecker product for assembling multiple-locus models in Chap. 3, we were dealing only with biallelic loci. As we have in the present chapter introduced multiallelic models, we shall hereafter provide an example of a model with multiple loci and multiple alleles, the simplest case of which is a two-locus model with a biallelic locus (locus A^1) and a three-allele locus (locus A^2). Since the dimension of the one-locus three-allele model is 6 (see, e.g., Expression 4.12), the dimension of the resulting two-locus model is $6 \times 3 = 18$. With G_{1111} as reference, the genetic-effects design matrix of such model can be obtained (from Expressions 4.5, 4.12) as:

$$\mathbf{S} = \mathbf{S}_2 \otimes \mathbf{S}_1$$

$$= \begin{pmatrix} 1 & 0 & 0 & 0 & 0 & 0 \\ 1 & 1 & 1 & 0 & 0 & 0 \\ 1 & 2 & 0 & 0 & 0 & 0 \\ 1 & 0 & 0 & 1 & 1 & 0 \\ 1 & 1 & 0 & 1 & 0 & 1 \\ 1 & 0 & 0 & 2 & 0 & 0 \end{pmatrix} \otimes \begin{pmatrix} 1 & 0 & 0 \\ 1 & 1 & 1 \\ 1 & 2 & 0 \end{pmatrix}, \tag{4.14}$$

and the model in the $\mathbf{G} = \mathbf{S}\ \mathbf{E}$ form can then be expressed as:

$$\begin{pmatrix} G_{1111} \\ G_{1211} \\ G_{2211} \\ G_{1112} \\ G_{1212} \\ G_{2212} \\ G_{1122} \\ G_{1222} \\ G_{2222} \\ G_{1113} \\ G_{1213} \\ G_{2213} \\ G_{1123} \\ G_{1223} \\ G_{2223} \\ G_{1133} \\ G_{1233} \\ G_{2233} \end{pmatrix} = \begin{pmatrix} 1 & 0 & 0 & 0 & 0 & 0 & 0 & 0 & 0 & 0 & 0 & 0 & 0 & 0 & 0 & 0 & 0 & 0 \\ 1 & 1 & 1 & 0 & 0 & 0 & 0 & 0 & 0 & 0 & 0 & 0 & 0 & 0 & 0 & 0 & 0 & 0 \\ 1 & 2 & 0 & 0 & 0 & 0 & 0 & 0 & 0 & 0 & 0 & 0 & 0 & 0 & 0 & 0 & 0 & 0 \\ 1 & 1 & 1 & 1 & 1 & 1 & 1 & 1 & 1 & 0 & 0 & 0 & 0 & 0 & 0 & 0 & 0 & 0 \\ 1 & 2 & 0 & 1 & 2 & 0 & 1 & 2 & 0 & 0 & 0 & 0 & 0 & 0 & 0 & 0 & 0 & 0 \\ 1 & 0 & 0 & 2 & 0 & 0 & 0 & 0 & 0 & 0 & 0 & 0 & 0 & 0 & 0 & 0 & 0 & 0 \\ 1 & 1 & 1 & 2 & 2 & 2 & 0 & 0 & 0 & 0 & 0 & 0 & 0 & 0 & 0 & 0 & 0 & 0 \\ 1 & 2 & 0 & 2 & 4 & 0 & 0 & 0 & 0 & 0 & 0 & 0 & 0 & 0 & 0 & 0 & 0 & 0 \\ 1 & 0 & 0 & 0 & 0 & 0 & 0 & 0 & 0 & 1 & 0 & 0 & 1 & 0 & 0 & 0 & 0 & 0 \\ 1 & 1 & 1 & 0 & 0 & 0 & 0 & 0 & 0 & 1 & 1 & 1 & 1 & 1 & 1 & 0 & 0 & 0 \\ 1 & 2 & 0 & 0 & 0 & 0 & 0 & 0 & 0 & 1 & 2 & 0 & 1 & 2 & 0 & 0 & 0 & 0 \\ 1 & 0 & 0 & 1 & 0 & 0 & 0 & 0 & 0 & 1 & 0 & 0 & 0 & 0 & 0 & 1 & 0 & 0 \\ 1 & 1 & 1 & 1 & 1 & 1 & 0 & 0 & 0 & 1 & 1 & 1 & 0 & 0 & 0 & 1 & 1 & 1 \\ 1 & 2 & 0 & 1 & 2 & 0 & 0 & 0 & 0 & 1 & 2 & 0 & 0 & 0 & 0 & 1 & 2 & 0 \\ 1 & 0 & 0 & 0 & 0 & 0 & 0 & 0 & 0 & 2 & 0 & 0 & 0 & 0 & 0 & 0 & 0 & 0 \\ 1 & 1 & 1 & 0 & 0 & 0 & 0 & 0 & 0 & 2 & 2 & 2 & 0 & 0 & 0 & 0 & 0 & 0 \\ 1 & 2 & 0 & 0 & 0 & 0 & 0 & 0 & 0 & 2 & 4 & 0 & 0 & 0 & 0 & 0 & 0 & 0 \end{pmatrix} \begin{pmatrix} R \\ a^{1} \\ d^{1} \\ a^{2.12} \\ aa^{12} \\ da^{12} \\ d^{2.12} \\ ad^{12} \\ dd^{12} \\ a^{2.13} \\ aa^{13} \\ da^{13} \\ d^{2.13} \\ ad^{13} \\ dd^{13} \\ d^{2.23} \\ ad^{23} \\ dd^{23} \end{pmatrix} \tag{4.15}$$

Note that we have added the indicators of the single loci to the superscripts of the genetic effects, as we have already done previously (e.g., Expressions 2.31 and 2.32). Note also that the genetic-effects design matrix of each new locus is added to the model by operating it (using the Kronecker product) at the left-hand side of the previous locus or loci involved (see Expression 4.14), as explained in the previous chapter.

The equivalent expression of the form $\mathbf{E} = \mathbf{S}^{-1}\ \mathbf{G}$ can be obtained either from Expressions 4.4 or 4.11—in a way analogous to Expression (4.14). It can of course also be obtained by inverting the 18×18 genetic-effects design matrix in Expression 4.15. Either way, we obtain:

$$\begin{pmatrix} R \\ a^{1} \\ d^{1} \\ a^{2.12} \\ aa^{12} \\ da^{12} \\ d^{2.12} \\ ad^{12} \\ dd^{12} \\ a^{2.13} \\ aa^{13} \\ da^{13} \\ d^{2.13} \\ ad^{13} \\ dd^{13} \\ d^{2.23} \\ ad^{23} \\ dd^{23} \end{pmatrix} = \begin{pmatrix}
1 & 0 & 0 & 0 & 0 & 0 & 0 & 0 & 0 & 0 & 0 & 0 & 0 & 0 & 0 & 0 & 0 & 0 \\
-\tfrac{1}{2} & 0 & \tfrac{1}{2} & 0 & 0 & 0 & 0 & 0 & 0 & 0 & 0 & 0 & 0 & 0 & 0 & 0 & 0 & 0 \\
-\tfrac{1}{2} & 1 & -\tfrac{1}{2} & 0 & 0 & 0 & 0 & 0 & 0 & 0 & 0 & 0 & 0 & 0 & 0 & 0 & 0 & 0 \\
-\tfrac{1}{2} & 0 & 0 & 0 & 0 & 0 & \tfrac{1}{2} & 0 & 0 & 0 & 0 & 0 & 0 & 0 & 0 & 0 & 0 & 0 \\
\tfrac{1}{4} & 0 & -\tfrac{1}{4} & 0 & 0 & 0 & -\tfrac{1}{4} & 0 & \tfrac{1}{4} & 0 & 0 & 0 & 0 & 0 & 0 & 0 & 0 & 0 \\
\tfrac{1}{4} & -\tfrac{1}{2} & \tfrac{1}{4} & 0 & 0 & 0 & -\tfrac{1}{4} & \tfrac{1}{2} & -\tfrac{1}{4} & 0 & 0 & 0 & 0 & 0 & 0 & 0 & 0 & 0 \\
-\tfrac{1}{2} & 0 & 0 & 1 & 0 & 0 & -\tfrac{1}{2} & 0 & 0 & 0 & 0 & 0 & 0 & 0 & 0 & 0 & 0 & 0 \\
\tfrac{1}{4} & 0 & -\tfrac{1}{4} & -\tfrac{1}{2} & 0 & \tfrac{1}{2} & \tfrac{1}{4} & 0 & -\tfrac{1}{4} & 0 & 0 & 0 & 0 & 0 & 0 & 0 & 0 & 0 \\
\tfrac{1}{4} & -\tfrac{1}{2} & \tfrac{1}{4} & -\tfrac{1}{2} & 1 & -\tfrac{1}{2} & \tfrac{1}{4} & \tfrac{1}{2} & 0 & 0 & 0 & 0 & 0 & 0 & 0 & 0 & 0 & \\
-\tfrac{1}{2} & 0 & 0 & 0 & 0 & 0 & 0 & 0 & 0 & 0 & 0 & 0 & 0 & 0 & 0 & \tfrac{1}{2} & 0 & 0 \\
\tfrac{1}{4} & 0 & -\tfrac{1}{4} & 0 & 0 & 0 & 0 & 0 & 0 & 0 & 0 & 0 & 0 & 0 & 0 & -\tfrac{1}{4} & 0 & \tfrac{1}{4} \\
\tfrac{1}{4} & -\tfrac{1}{2} & \tfrac{1}{4} & 0 & 0 & 0 & 0 & 0 & 0 & 0 & 0 & 0 & 0 & 0 & 0 & -\tfrac{1}{4} & \tfrac{1}{2} & -\tfrac{1}{4} \\
-\tfrac{1}{2} & 0 & 0 & 1 & 0 & 0 & 0 & 0 & 0 & 1 & 0 & 0 & 0 & 0 & 0 & -\tfrac{1}{2} & 0 & 0 \\
\tfrac{1}{4} & 0 & -\tfrac{1}{4} & 1 & 0 & 0 & 0 & 0 & 0 & -\tfrac{1}{2} & 0 & \tfrac{1}{2} & 0 & 0 & 0 & \tfrac{1}{4} & 0 & -\tfrac{1}{4} \\
\tfrac{1}{4} & -\tfrac{1}{2} & \tfrac{1}{4} & 1 & 0 & 0 & 0 & 0 & 0 & -\tfrac{1}{2} & 1 & -\tfrac{1}{2} & 0 & 0 & 0 & \tfrac{1}{4} & -\tfrac{1}{2} & \tfrac{1}{4} \\
0 & 0 & 0 & 0 & 0 & 0 & -\tfrac{1}{2} & 0 & 0 & 0 & 0 & 0 & 0 & 0 & 0 & -\tfrac{1}{2} & 0 & 0 \\
0 & 0 & 0 & 0 & 0 & 0 & 0 & 0 & -\tfrac{1}{4} & 0 & 0 & 0 & -\tfrac{1}{2} & 0 & \tfrac{1}{2} & \tfrac{1}{4} & 0 & -\tfrac{1}{4} \\
0 & 0 & 0 & 0 & 0 & 0 & 0 & -\tfrac{1}{2} & \tfrac{1}{4} & 0 & 0 & 0 & -\tfrac{1}{2} & 1 & -\tfrac{1}{2} & \tfrac{1}{4} & -\tfrac{1}{2} & \tfrac{1}{4}
\end{pmatrix}$$

$$\times \begin{pmatrix} G_{1111} \\ G_{1211} \\ G_{2211} \\ G_{1112} \\ G_{1212} \\ G_{2212} \\ G_{1122} \\ G_{1222} \\ G_{2222} \\ G_{1113} \\ G_{1213} \\ G_{2213} \\ G_{1123} \\ G_{1223} \\ G_{2223} \\ G_{1133} \\ G_{1233} \\ G_{2233} \end{pmatrix} \tag{4.16}$$

This has been just a taste of how inconvenient to handle on paper the models become already under relatively simple situations (in this case, two loci with five alleles in total). Some readers may find it interesting to track which genetic effects are involved in obtaining each genotypic value from the reference of G_{1111} in Expression 4.14 and/or in having a look at how the genetic effects are defined in terms of the genotypic values in Expression 4.15. Note, for instance, that all marginal effects are defined in the same way as in the biallelic models only (in terms of the genotypic values implied in the particular allele substitution they stand for).

Fortunately, in any case, we will in general not be interested in making the genetic models explicit. Instead, we are interested in knowing the precise way to tell a computer how to handle them and provide us with the results we aim (e.g., whether there are significant epistatic interactions between two loci). The way described above to build multiple-locus models from single-locus models can also be applied with models accounting for other important genetic circumstances, as the ones we address in what follows.

4.7 Gene-Environment Interaction

In Chap. 2, we have considered that the environment would affect the individuals randomly, in particular, regardless of their genotypes. The environment was thus a nuisance effect that simply obscured to whatever extent the underlying genetic influence on the phenotype—like a noise that prevents us from hearing properly or a veil that distorts certain details of the objects we intend to observe. But, as introduced from the beginning of this book, the action of the environment may also fit a pattern, in which case a more complex GP map—which strictly speaking we could call a genotype-and-environment-to-phenotype map—shall be required to properly model the trait.

In some cases, the influence of an environmental factor (e.g., average temperature) on the trait may follow a very easy-to-track pattern. For instance, one alternative of the environmental factor (e.g., an average temperature of 20 °C) may systematically increase all genotypic values in the same way (e.g., plus 3 cm) as compared to a different alternative of the environmental factor (e.g., an average temperature of 15 °C). In this case, we could simply "correct" the data by removing the difference between the two environmental alternatives (3 cm) from the individuals at 20 °C and thus avoid having to increase the complexity of the models for accounting for the d ifference between the two environments.

However, extremely interesting cases exist that follow more complex patterns and thus require a GP map accounting for systematic effects of environmental variables in order not to miss key facts of the system under study. The case of two alleles at a locus displaying higher phenotypes under (or being better adapted to) two different alternatives of an environmental factor (see Fig. 4.4) is indeed one such case. Each environmental factor imposing a pattern on the system may then be modeled through a matrix equivalent to a genetic-effects design matrix—an environmental-effects design matrix. Along these lines, Ma et al. (2012) described the influence of an environmental factor with two alternatives (particularly, smokers vs. non-smokers) in the form $\mathbf{G} = \mathbf{SE}$ as:

$$\begin{pmatrix} G_1 \\ G_2 \end{pmatrix} = \begin{pmatrix} 1 & 0 \\ 1 & 1 \end{pmatrix} \begin{pmatrix} R \\ e \end{pmatrix}, \tag{4.17}$$

or, equivalently, in the form $\mathbf{E} = \mathbf{S}^{-1}\mathbf{G}$ as:

$$\begin{pmatrix} R \\ e \end{pmatrix} = \begin{pmatrix} 1 & 0 \\ -1 & 1 \end{pmatrix} \begin{pmatrix} G_1 \\ G_2 \end{pmatrix}. \tag{4.18}$$

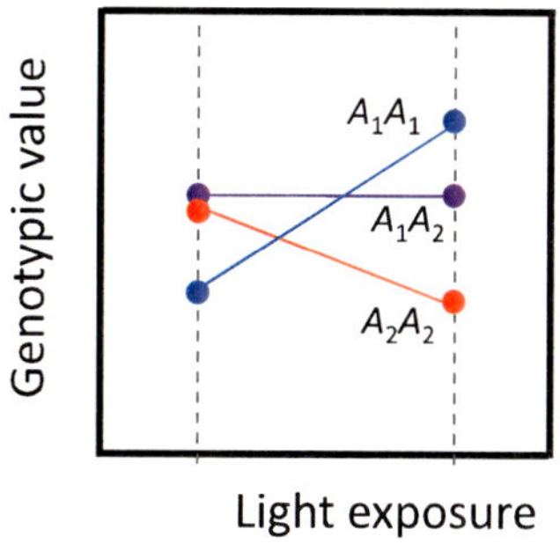

Fig. 4.4 Gene-by-environment interactions. Biallelic locus A has different genotypic values for different light exposures. Each allele is adapted to a different exposure while the heterozygote performs relatively well under both environmental conditions. There is gene-environment interaction whenever the lines (the reaction norms) are not parallel. See a similar example in Álvarez-Castro (2016)

Expression 4.17 shows that a certain genotypic value, G, in the environment taken as reference (environment E^1), G_1, plus an environmental effect, e, provides the genotypic value at a second environment (environment E^2), G_2. Equivalently, Expression 4.18 shows that the environmental effect is defined as $e = G_2 - G_1$. It is noteworthy that these expressions also fit a haploid, biallelic locus. In any case, we can here generalize this expression to multiple (in particular, m) possible alternatives of an environmental variable simply by defining the environmental effects $e^{1i} = G_i - G_1$, $i = 2, \ldots, m$. Indeed, any other environmental effect, e_{ij}, $i \neq 1$, can then be obtained from them as $e^{ij} = e^{1j} - e^{1i}$. In accordance with these definitions, we can extend Expressions 4.17, 4.18 to, for instance, $m = 3$, as:

$$\begin{pmatrix} G_1 \\ G_2 \\ G_3 \end{pmatrix} = \begin{pmatrix} 1 & 0 & 0 \\ 1 & 1 & 0 \\ 1 & 0 & 1 \end{pmatrix} \begin{pmatrix} R \\ e^{12} \\ e^{13} \end{pmatrix}, \tag{4.19}$$

or, equivalently:

$$\begin{pmatrix} R \\ e^{12} \\ e^{13} \end{pmatrix} = \begin{pmatrix} 1 & 0 & 0 \\ -1 & 1 & 0 \\ -1 & 0 & 1 \end{pmatrix} \begin{pmatrix} G_1 \\ G_2 \\ G_3 \end{pmatrix}. \tag{4.20}$$

Expressions 4.19 and 4.20 enable, for instance, to consider separate environmental effects for smokers, passive smokers, and strictly non-smokers.

In any case (i.e., for whatever number of possible values one environmental factor may take), we shall be particularly interested in knowing whether the patterns the environments impose to the genotypic values of the different genotypes are different, as in Fig. 4.4—i.e., whether they interact. This requires to implement the model with parameters that account for gene-environment interaction. Such parameters arise naturally using the Kronecker product, just as the epistasis parameters did in Expressions 3.3 and 3.4 and related text in Chap. 3 and in Expressions 4.14–4.16.

Hereafter we illustrate the simplest case of a one-locus two-allele system with two different instances for an environmental factor. For this case, the genetic-environment-effects design matrix can be obtained using the Kronecker product (in the reverse order) of the genetic-effects design matrix and the environment-effects design matrix in Expressions 4.5 and 4.17 as:

$$\mathbf{S} = \begin{pmatrix} 1 & 0 \\ 1 & 1 \end{pmatrix} \otimes \begin{pmatrix} 1 & 0 & 0 \\ 1 & 1 & 1 \\ 1 & 2 & 0 \end{pmatrix}, \tag{4.21}$$

leading to the corresponding expression $\mathbf{G} = \mathbf{S}\ \mathbf{E}$, expanding to:

$$\begin{pmatrix} G_{111} \\ G_{121} \\ G_{221} \\ G_{112} \\ G_{122} \\ G_{222} \end{pmatrix} = \begin{pmatrix} 1 & 0 & 0 & 0 & 0 & 0 \\ 1 & 1 & 1 & 0 & 0 & 0 \\ 1 & 2 & 0 & 0 & 0 & 0 \\ 1 & 0 & 0 & 1 & 0 & 0 \\ 1 & 1 & 1 & 1 & 1 & 1 \\ 1 & 2 & 0 & 1 & 2 & 0 \end{pmatrix} \begin{pmatrix} R \\ a \\ d \\ e \\ ae \\ de \end{pmatrix}, \tag{4.22}$$

where the third digit of the subscript of each genotypic value indicates the environment and *ae* and *de* are the gene-environment interaction parameters—the additive-environment and the dominance-environment parameters, respectively. For $ae = de = 0$, the lines in Fig. 4.4 would be parallel. Expression 4.22 was first provided by Ma et al. (2012), although using a slightly different notation.

The Kronecker product naturally enables considering more than one environmental factor in the same model, like temperature and humidity. To do so, we just have to add one environmental-effects design matrix for each environmental factor. Analogously, more loci can be added to the model by just adding one genetic-effects design matrix for each of them, as explained in the multiple-locus section above. It is of course important not to confuse to implement multiple possible environmental values within one environmental factor (as considered in Expressions 4.19 and 4.20 above) with to implement several environmental factors (using the Kronecker product of the environmental-effects design matrices of the different environmental factors). The first case is analogous to considering several alleles in one locus, whereas the second one is analogous to considering several loci.

4.8 Sex-Linked Loci

Not having made it explicit otherwise, we have hitherto considered only autosomal loci. The present section is devoted to sex-linked loci. We shall start with loci at the chromosome determining the homogametic sex, which we shall refer to as X-linked inheritance, in relation with the XY sex-determination system. Note that, nonetheless, these developments also fit the Z-linked inheritance of the ZW sex-determination system. Then, we shall move to Y-linked (also valid for W-linked) inheritance. Note also that pseudoautosomal loci (those present in the two sex chromosomes) fit to the models for autosomal loci.

4.8.1 X-Linked Loci

At a two-allele X-linked locus (with alleles X_1, X_2), individuals cannot only be homozygous (X_1X_1 and X_2X_2) or heterozygous (X_1X_2), which happens in females, but also hemizygous (X_1Y and X_2Y), which affects males. Each of those genotypes may code for a potentially different phenotype (G_{11}, G_{22}, G_{12}, G_1 and G_2, respectively). In particular, the genetic system may potentially lead to different phenotypes in males and females. Besides, each sex has independent sets of allele substitutions. Therefore, we may naturally model a different reference point for each sex, R_f and R_m. Then, the female genetic effects are additive and dominant, a_f and d_f, since interaction may arise in the heterozygote, whereas in males we can only have an additive genetic effect, a_m. Thus both the vector of genotypic values and the vector of genetic effects have six scalars.

Concerning now the 6 × 6 square genetic-effects design matrix, **S**, it could be built using matrices we have already developed, particularly in Expressions 4.5 and 4.17, leading to:

$$\begin{pmatrix} G_{11} \\ G_{12} \\ G_{22} \\ G_1 \\ G_2 \end{pmatrix} = \begin{pmatrix} 1 & 0 & 0 & 0 & 0 \\ 1 & 1 & 1 & 0 & 0 \\ 1 & 2 & 0 & 0 & 0 \\ 0 & 0 & 0 & 1 & 0 \\ 0 & 0 & 0 & 1 & 1 \end{pmatrix} \begin{pmatrix} R_f \\ a_f \\ d_f \\ R_m \\ a_m \end{pmatrix}. \tag{4.23}$$

Indeed, the resulting matrix is a block-diagonal matrix with two blocks. The upper-left 3 × 3 block of the matrix in this expression reflects that the system behaves as a one-locus two-allele autosomal locus for females (Expression 4.5). In what regards the males, the system is not diploid any longer, but equivalent to a haploid system, and this is why the 2 × 2 bottom-right block is the same as that of an environmental factor (Expression 4.17).

The upper-right and lower-left blocks of zeros of the genetic-effects design matrix in Expression 4.23 reflect that the system is set so that males and females are considered independently. Indeed, its inverse matrix can be computed from the inverses of the (remaining) diagonal blocks. Thus, from Expressions 4.4 and 4.18, it is easy to derive that the equivalent form of Expression 4.23 is:

$$\begin{pmatrix} R_f \\ a_f \\ d_f \\ R_m \\ a_m \end{pmatrix} = \begin{pmatrix} 1 & 0 & 0 & 0 & 0 & 0 \\ -\frac{1}{2} & 0 & \frac{1}{2} & 0 & 0 & 0 \\ -\frac{1}{2} & 1 & -\frac{1}{2} & 0 & 0 & 0 \\ 0 & 0 & 0 & 0 & 1 & 0 \\ 0 & 0 & 0 & 0 & -1 & 1 \end{pmatrix} \begin{pmatrix} G_{11} \\ G_{12} \\ G_{22} \\ G_1 \\ G_2 \end{pmatrix}. \qquad (4.24)$$

The extension of Expressions 4.23 and 4.24 to multiple alleles is trivial—by just using the corresponding multiallelic expressions of the upper-left and lower-right blocks of the matrices, from Expressions 4.11, 4.12, 4.19, and 4.20 and related text.

This is not the only way in which a model of an X-linked locus can be developed. We could alternatively consider the by-default hypothesis that the two sexes may display equal phenotypes. This is indeed the case of the trait at which this kind of inheritance was first described—the white eyes mutation of *Drosophila melanogaster* (Morgan 1910)—incidentally, at the early stages of the field of genetics. It may thus be considered convenient that the model implements parameters indicating whether those potential differences between the sexes actually occur.

We may therefore be interested in a model that directly accounts for the potential differences in phenotype between the hemizygous genotypes (thus, in males) and their corresponding homozygous genotypes (thus, in females). Such parameters would simply equal zero whenever those two pairs of genotypes (X_1X_1 and X_1Y, and X_2X_2, and X_2Y) display a common phenotype—i.e., when $G_{11} = G_1$ and $G_{22} = G_2$. Such a model can be developed by just modifying the two last rows of Expression 4.23, as:

$$\begin{pmatrix} G_{11} \\ G_{12} \\ G_{22} \\ G_1 \\ G_2 \end{pmatrix} = \begin{pmatrix} 1 & 0 & 0 & 0 & 0 \\ 1 & 1 & 1 & 0 & 0 \\ 1 & 2 & 0 & 0 & 0 \\ 1 & 0 & 0 & 1 & 0 \\ 1 & 2 & 0 & 0 & 1 \end{pmatrix} \begin{pmatrix} R \\ a \\ d \\ x_1 \\ x_2 \end{pmatrix}. \qquad (4.25)$$

Actually, we have also changed the notation of the variables of the vector of genetic effects. This is in accordance with the fact that the male genetic effects have now a different biological interpretation, as further explained below. By inverting the genetic effects matrix in Expression 4.25, we get to its equivalent form:

$$\begin{pmatrix} R \\ a \\ d \\ x_1 \\ x_2 \end{pmatrix} = \begin{pmatrix} 1 & 0 & 0 & 0 & 0 \\ -\frac{1}{2} & 0 & \frac{1}{2} & 0 & 0 \\ -\frac{1}{2} & 1 & -\frac{1}{2} & 0 & 0 \\ -1 & 0 & 0 & 1 & 0 \\ 0 & 0 & -1 & 0 & 1 \end{pmatrix} \begin{pmatrix} G_{11} \\ G_{12} \\ G_{22} \\ G_1 \\ G_2 \end{pmatrix}. \tag{4.26}$$

The extension of Expressions 4.25 and 4.26 to several alleles can be done either from Expression 4.25 or from Expression 4.26 in an easy way. For instance, a k-allele expression analogous to Expression 4.26 can be obtained using the corresponding autosomal expression as the upper-left block, followed by a block of zeros to the right, and k rows below defining the k parameters of the departures between hemizygous and homozygous genotypes (by placing the 1 and the -1 in the correct places and zeros otherwise).

In Expression 4.23, the penultimate row of the genetic-effects design matrix just sets the male reference point to be G_1, as the first row of it inverse matrix in Expression 4.24 confirms. Concerning now the last row of the matrix in Expression 4.23, its penultimate scalar indicates that we start the measure at that reference point (G_1), while the last one indicates that we add up a quantity measured by $a_♂$ for moving from the reference genotypic value (G_1) to the other genotypic value (G_2). Again, this can be confirmed through Expression 4.24, whose last row indicates that $a_♂ = G_2 - G_1$. Thus, $a_♂$ is just meant to measure the effect of making an allele substitution from X_1 to X_2 in a male.

In Expression 4.25, the first scalar of each of the two last rows of the genetic-effects design matrix indicates that we keep on using the same reference point as in females, G_{11}. In the penultimate row, the penultimate scalar indicates that x_1 measures the difference between that reference point and G_1. Indeed, the penultimate row is equal to the first row but for that penultimate scalar. Analogously, the last row is equal to the third row but for the last scalar, which simply indicates that x_2 measures the difference between G_{22} and G_2. Indeed, Expression 4.26 shows that these parameters are defined as $x_i = G_i - G_{ii}$, $i = 1, 2$. Therefore, x_1 and x_2 indeed measure departures from the by-default values the hemizygous genotypes may display—those of their corresponding homozygous genotypes. Figure 4.5 shows a graphical interpretations of the genetic effects in Expressions 4.25 and 4.26.

Overall, neither Expressions 4.23 and 4.24 nor Expressions 4.25 and 4.26 are wrong—they just focus more directly on different aspects of the genetic system. Expressions 4.23 and 4.24 are a good choice if we assume that all genotypic values may potentially be different and they use male and female independent sets of genetic effects for each of them, all of which fit the conventional way the genetic effects are defined for autosomal loci. Since the interpretation of the reference points and the additive genetic effects of males and females are equivalent for the two sexes, we use the same notation for them, just indicating the sex with a subscript of all variables in order to avoid confusion.

Expressions 4.25 and 4.26, on the other hand, describe the genotypic values of the (hemizygous) male genotypes in terms of how they differ from their corresponding

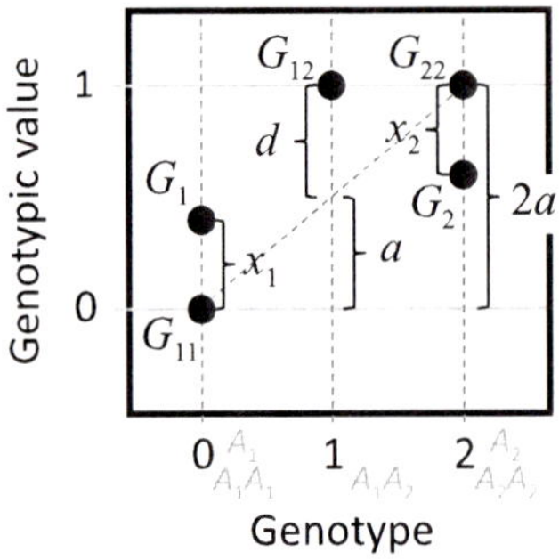

Fig. 4.5 Graphical representation of the individual-referenced genetic effects of an X-linked locus, with the male genetic effects being set as departures from the females. The genotypes are in the horizontal axis represented as content of allele A_2 (in the hemizygous genotypes the alleles are worth two) and the genotypic values (G_{ij} for females and G_i for males) are in the vertical axis. The difference between the genotypic values of the two female homozygotes determines the additive effect (of substituting allele A_1 by allele A_2) and the female dominance effect is the departure of the heterozygote from its additive expectation (due to the presence of interaction between the two alleles), just as in Fig. 4.1. The male genetic effects, x_i, are conceived as departures of the hemizygous (male) genotypic values from the homozygous (female) genotypic values

(homozygous) female genotypes. These expressions provide parameters that directly show whether and how much the hemizygous genotypes and their corresponding homozygous genotypes differ, if at all. Whereas the female genetic effects of these expressions are defined in the conventional way (as in autosomal loci), the male genetic effects measure departures, which makes special sense, for instance, when we suspect that they may actually coincide. Thus, the nomenclature for males and females is different, and there is no need for gender-related subscripts.

4.8.2 Y-Linked Loci

A Y-linked locus is basically a haploid locus. Thus, in principle, in can be modeled by just adapting the notation of Expressions 4.17–4.20, developed above for environmental factors (in particular, an environmental effect, e, in those expressions turns to be an additive effect, a, in the context of a Y-linked loci). As opposed to an environmental factor, a Y-linked locus is a genetic locus by itself, and thus Expressions 4.17 and 4.20 constitute a meaningful model—a one-Y-linked-locus model.

Expressions 4.21 and 4.22, developed above for integrating an environmental factor into a model of genetic—and environmental—effects, could also make sense in the context of a Y-linked locus (by just modifying the notation in the sense mentioned right above). In particular, they fit a two-locus model with an autosomal locus a Y-linked locus. However, adding up an autosomal locus to the model rises the issue of females also being considered, since they may display a trait that is partially affected (only in males) by a Y-linked locus. When this is the case,

Expressions 4.17–4.20 have to be implemented to accommodate a female genotypic value, as follows:

$$\begin{pmatrix} G_1 \\ G_2 \\ G_{♀} \end{pmatrix} = \begin{pmatrix} 1 & 0 & 0 \\ 1 & 1 & 0 \\ 1 & ½ & 1 \end{pmatrix} \begin{pmatrix} R \\ a \\ f \end{pmatrix}, \tag{4.27}$$

or, equivalently:

$$\begin{pmatrix} R \\ a \\ f \end{pmatrix} = \begin{pmatrix} 1 & 0 & 0 \\ -1 & 1 & 0 \\ -½ & -½ & 1 \end{pmatrix} \begin{pmatrix} G_1 \\ G_2 \\ G_{♀} \end{pmatrix}. \tag{4.28}$$

There is one only possible female genotypic value, f, in what regards the influence of a Y-linked locus to a trait, since the only possible female genotype at that locus actually is not having a genotype at all. We here just provide explicitly the expression of a one-locus two-allele Y-linked locus. The developments provided in the section on several loci above are valid for implementing such locus into a multilocus setting. The extension of Expressions 4.27 and 4.28 to multiple alleles follows the same logic as the extension to multiple environments above, by just adding up to it one more dimension (row and column) for the female genotypic value and genetic effect. All scalars of that column are zeros except from the last one, which is one and coincides with the last scalar of the row. All remaining scalars of the row are $-1/k$, where k is the number of alleles.

4.9 Gene-Sex Interaction

Above we have considered that a locus, even when sex-linked, may display the same phenotypes in females and males. Now, the opposite is also possible—an autosomal locus may display different effects in the two sexes, which can be labeled as gene-sex interaction. This situation can be modeled analogously to Expressions 4.23 and 4.24 for an X-linked locus, although in this case the two non-nil blocks of the genetic-effects design matrix work in the very same way and are therefore equal:

$$\begin{pmatrix} G_{f11} \\ G_{f12} \\ G_{f22} \\ G_{m11} \\ G_{m12} \\ G_{m22} \end{pmatrix} = \begin{pmatrix} 1 & 0 & 0 & 0 & 0 & 0 \\ 1 & 1 & 1 & 0 & 0 & 0 \\ 1 & 2 & 0 & 0 & 0 & 0 \\ 0 & 0 & 0 & 1 & 0 & 0 \\ 0 & 0 & 0 & 1 & 1 & 1 \\ 0 & 0 & 0 & 1 & 2 & 0 \end{pmatrix} \begin{pmatrix} R_f \\ a_f \\ d_f \\ R_m \\ a_m \\ d_m \end{pmatrix}. \tag{4.29}$$

For equating the vector of genetic effects, the inverse of that genetic-effects design matrix can be obtained by just replacing the upper-left and lower-right matrices by their inverses, which can be taken from Expression 4.4.

Then again, as for the case of sex-linked loci above (Expressions 4.25, 4.26), we can provide an alternative expression in which the effects on one sex are described as departures from the effects on the other sex. For the case of gene-sex interaction, such model is even more obvious as it can be developed straightly through an interaction between sex and the locus in question. Since sex can be modeled in the very same mathematical way as an environmental factor (Expression 4.21), the genetic-effects design matrix can be obtained as in Expression 4.22. Thus, by just adapting the notation, we obtain:

$$\begin{pmatrix} G_{f11} \\ G_{f12} \\ G_{f22} \\ G_{m11} \\ G_{m12} \\ G_{m22} \end{pmatrix} = \begin{pmatrix} 1 & 0 & 0 & 0 & 0 & 0 \\ 1 & 1 & 1 & 0 & 0 & 0 \\ 1 & 2 & 0 & 0 & 0 & 0 \\ 1 & 0 & 0 & 1 & 0 & 0 \\ 1 & 1 & 1 & 1 & 1 & 1 \\ 1 & 2 & 0 & 1 & 2 & 0 \end{pmatrix} \begin{pmatrix} R \\ a \\ d \\ s \\ as \\ ds \end{pmatrix}, \tag{4.30}$$

where the reference point and additive and dominance effects refer to females and the sex effect accounting for the departures between the two sexes, s, implies also an additive-by-sex and a dominance-by-sex effects—as and ds, respectively.

The definition of the genetic effects can be more directly tracked using the equivalent form of Expression 4.30:

$$\begin{pmatrix} R \\ a \\ d \\ s \\ as \\ ds \end{pmatrix} = \begin{pmatrix} 1 & 0 & 0 & 0 & 0 & 0 \\ -\frac{1}{2} & 0 & \frac{1}{2} & 0 & 0 & 0 \\ -\frac{1}{2} & 1 & -\frac{1}{2} & 0 & 0 & 0 \\ -1 & 0 & 0 & 1 & 0 & 0 \\ \frac{1}{2} & 0 & -\frac{1}{2} & -\frac{1}{2} & 0 & \frac{1}{2} \\ \frac{1}{2} & -1 & \frac{1}{2} & -\frac{1}{2} & 1 & -\frac{1}{2} \end{pmatrix} \begin{pmatrix} G_{f11} \\ G_{f12} \\ G_{f22} \\ G_{m11} \\ G_{m12} \\ G_{m22} \end{pmatrix}, \quad (4.31)$$

where the reference point and the additive and dominance effects are female effects—no need for subscripts since there are no male genetic effects any longer. Expression 4.31 directly leads to, for instance, $as = \frac{G_{m22} - G_{m11}}{2} - \frac{G_{f22} - G_{f11}}{2}$, which actually is the difference between the male and the female additive effects, as we have set since Expression 4.2: $as = a_m - a_f$. Incidentally, this model, from the reference point of G_{f11}, provides an explicit parameter for the additive female effect, $a_f = a$, but it does not provide an explicit parameter for the additive male effect. Nevertheless, such parameter can be easily derived as $a_m = as + a$ whenever necessary. Alternatively, we could express the model from the reference of G_{m11}, in which case $a_m = a$.

The genetic-effects design matrix in Expression 4.31 is also the inverse matrix of the gene-environment interaction model Expression 4.22. As in previous cases, the extension to multiple alleles can be done by substituting biallelic blocks in the matrices by their corresponding multiallelic blocks. The extension to multiple loci can be done following the same logic as for the simpler one-locus models (see the section on multiple loci above).

4.10 Imprinting

Imprinting effects (also called parent-of-origin effects) occur when the effect of an allele in an individual is conditioned by the sex of the parent from which the allele came (assumingly because of different methylation patterns in the two sexes). Developing an individual-referenced model of imprinting is challenging and, because of that, also very illustrative about the development and interpretation of individual-referenced models in general. The following text is an adaptation of the development of an individual-referenced model of imprinting by Álvarez-Castro (2014):

When considering one imprinted locus with two alleles, we could be tempted to try to fit it into a one-locus four-allele genetic model, since each of the two alleles (with different nucleotide sequences) may be expressed at the level of the phenotype in two ways (each has two possible methylation stages), thus leading to a total of four variants with potentially different effects on the phenotype. However, it is not possible to use the multiallelic model for depicting the differences between phenotypes due to allelic variants, as explained below.

Let the two alleles be A_1 and A_2, just as in the cases without imprinting above. Due to imprinting, there now also exist the modified variants $\bar{A}_1$ and $\bar{A}_2$, summing up to a total of four variants as mentioned just above. In a four-allele model of genetic effects, there are six additive effects, three of which can be retrieved from the other three (see the multiallelic section above and Álvarez-Castro and Yang 2011). These parameters account for effects of allele substitutions between any possible pair of homozygotes, which in our case would be A_1A_1, A_2A_2, $\bar{A}_1\bar{A}_1$, and $\bar{A}_2\bar{A}_2$.

However, none of those genotypes will be present in any of the individuals of our analyses. More to the point, we cannot easily think of those genotypes as putative artificial constructs, since imprinted loci preclude viability under unbalanced dosages of modified alleles (Kono et al. 2004; Kawahara et al. 2007). Indeed, the two "homozygotes" of our imprinted biallelic locus actually are $A_1\bar{A}_1$ and $A_2\bar{A}_2$—they are allele-wise homozygotes, even if they are not variant-wise homozygotes. Then, we can measure the additive effect between those allele-wise homozygotes in a way analogous to a non-imprinted locus as $a = (G_{22} - G_{11})/2$ (Expression 4.2).

Thus, properly conceptualizing the additive effects of an imprinted locus may require some reflection, but they in the end can be modelled in a way that brings no additional complexity as compared to modelling the non-imprinted case. It is the modelling of the dominance effects that will make the difference. Indeed, because $A_1\bar{A}_1$ and $A_2\bar{A}_2$ are not variant-wise homozygotes, there is more than one possible allele substitution between them, which will enable several options to model the interaction effects, as shown right below.

4.10.1 Two Dominance Effects

Let us, for instance, start from the reference of $A_1\bar{A}_1$. Note that there are, indeed, two possible ways of performing one only allele substitution from this reference—leading to either $A_1\bar{A}_2$ or $A_2\bar{A}_1$. Consequently, considering two possible dominance effects (one for each parent-of-origin of the two alleles in the heterozygote) emerges as a sensible solution.

To begin with the development of this two-dominance setting, an expression of the genotypic values as a sum of genetic effects of allele substitutions from one reference genotype is firstly provided—as it was done in Expression 4.5 for a non-imprinted locus. Let G_{12} and G_{21} be the different (in case of imprinting) genotypic values of genotypes $A_1\bar{A}_2$ and $A_2\bar{A}_1$, respectively. Then, following the same logic as in the previous cases considered, the individual-referenced expression from the reference of homozygote $A_1\bar{A}_1$ can be obtained as:

$$\begin{pmatrix} G_{11} \\ G_{12} \\ G_{21} \\ G_{22} \end{pmatrix} = \begin{pmatrix} 1 & 0 & 0 & 0 \\ 1 & 1 & 1 & 0 \\ 1 & 1 & 0 & 1 \\ 1 & 2 & 0 & 0 \end{pmatrix} \begin{pmatrix} R \\ a \\ d^{12} \\ d^{21} \end{pmatrix}. \tag{4.32}$$

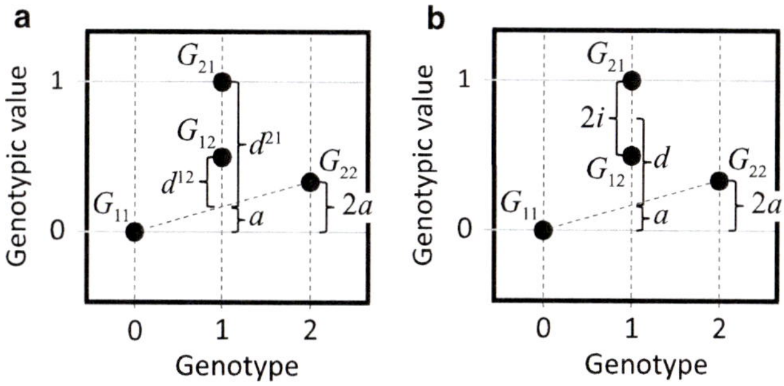

Fig. 4.6 Individual-referenced genetic effects proposed in the text for a one-locus, two-allele imprinted genetic system. The alleles are A_1 and A_2, with the variants due to imprinting being $\bar{A}_1$ and $\bar{A}_2$. (**a**) The two-dominances model is a natural extension of the non-imprinted case that consists in introducing two dominance parameters, one for each heterozygote. As well as in the non-imprinted case, the dominance effects measure departures of the heterozygotes from their additive expectation (midpoint between the homozygotes). (**b**) The imprinting-effect model keeps one only dominance effect that accounts for the departure of the midpoint between the two heterozygotes and the midpoint between the homozygotes and adds up an imprinting effect for accounting for the distance between the heterozygotes and their midpoint. Thus, the dominance effect in this model coincides with that of a non-imprinted model (i.e., the one that would be obtained if imprinting was just disregarded)

The genotypic value of $A_2\bar{A}_2$ is here expressed as the sum of two additive effects from the reference, while the genotypic values of the heterozygotes involve one additive plus one dominance effect each. The difference between Expressions 4.32 and 4.5 is that in Expression 4.32, each heterozygote involves a different dominance effect. By equating the vector of genetic effects in Expression 4.32, we obtain an extension of Expression 4.4 to imprinting, providing how each of the genetic effects is defined in terms of the genotypic values:

$$\begin{pmatrix} R \\ a \\ d^{12} \\ d^{21} \end{pmatrix} = \begin{pmatrix} 1 & 0 & 0 & 0 \\ -½ & 0 & 0 & ½ \\ -½ & 1 & 0 & -½ \\ -½ & 0 & 1 & -½ \end{pmatrix} \begin{pmatrix} G_{11} \\ G_{12} \\ G_{21} \\ G_{22} \end{pmatrix}. \tag{4.33}$$

Thus, for instance, the second dominance effect is defined as $d^{21} = G_{21} - ½(G_{11} + G_{22})$.

4.10.2 Explicit Imprinting Effect

The previous setting can be used for detecting imprinting by just developing a procedure for testing whether the two dominance effects are significantly different.

To this aim, it seems nevertheless more convenient to design a model in which a parameter accounts for the difference between the two heterozygotes, thus leading to a more direct test for imprinting—consisting in just checking whether that parameter is significantly different from zero. Actually, this is in general terms the approach commonly chosen to model imprinting (see e.g. Mantey et al. 2005; Wolf et al. 2008). Hereafter, an individual-referenced model of imprinting is developed that implements a parameter to account for the putative difference between the heterozygotes with different parent-of-origin (i.e., an imprinting effect). As in the former option (Expressions 4.32 and 4.33), an expression of effects of allele substitutions from the reference of homozygote $A_1\bar{A}_1$ is here provided in the first place, as:

$$\begin{pmatrix} G_{11} \\ G_{12} \\ G_{21} \\ G_{22} \end{pmatrix} = \begin{pmatrix} 1 & 0 & 0 & 0 \\ 1 & 1 & 1 & -1 \\ 1 & 1 & 1 & 1 \\ 1 & 2 & 0 & 0 \end{pmatrix} \begin{pmatrix} R \\ a \\ d \\ i \end{pmatrix}. \tag{4.34}$$

This model is designed for using the midpoint between the two heterozygotes to define the dominance effect and the deviations of the two heterozygotes from that point as the imprinting effect. A graphical comparison of the two ways proposed here for modelling imprinting—the two-dominances model and the model with an explicit imprinting effect—is shown in Fig. 4.6. By equating the vector of genetic effects in Expression 4.34, it follows:

$$\begin{pmatrix} R \\ a \\ d \\ i \end{pmatrix} = \begin{pmatrix} 1 & 0 & 0 & 0 \\ -\frac{1}{2} & 0 & 0 & \frac{1}{2} \\ -\frac{1}{2} & \frac{1}{2} & \frac{1}{2} & -\frac{1}{2} \\ 0 & -\frac{1}{2} & \frac{1}{2} & 0 \end{pmatrix} \begin{pmatrix} G_{11} \\ G_{12} \\ G_{21} \\ G_{22} \end{pmatrix}. \tag{4.35}$$

From this expression, it immediately follows that indeed $d = \frac{1}{2}(G_{12} + G_{21}) - \frac{1}{2}(G_{11} + G_{22})$ (i.e., the dominance effect measures the distance of the midpoint between the two heterozygotes and the additive expectation) and $i = \frac{1}{2}(G_{21} - G_{12})$ (i.e., the imprinting effect measures the distance of the heterozygotes from the midpoint between them). Expressions 4.34 and 4.35 provide a general individual-referenced formulation with an explicit imprinting effect, analogously to Expressions 4.32 and 4.33 for the two-dominance model above. As in previous models, the definition of the genetic effects facilitates an extension to multiple alleles via the inverse of the genetic-effects design matrix—i.e., extensions of Expressions 4.33 and 4.35.

4.11 The Effect of a Gene

If someone completely unaware of the basics of quantitative genetics would try to guess what the expression "the effect of a gene" means, that person could easily find it plausible that it refers to what happens when that gene is present as compared to when it is not—i.e., the effect of knocking out a gene. However, in the quantitative genetics context, a gene with no effect on a phenotype is one that shows no correlation between its genetic variability and the phenotypic variation. Interestingly, this may happen just because the gene in question does not display any genetic variability at all within the range of the study.

Indeed, it may look paradoxical to the layperson that a gene having a key role in a molecular route of the physiology of a phenotype may not have an effect on the trait (in the context of quantitative genetics) as long as there is no genetic variability at that gene (for the individuals under study). It might even happen that a gene nuclear for a trait undergoes so strong purifying selection at a population that any significant variability is efficiently removed, thus not having an effect on the trait in terms of quantitative genetics.

4.11.1 Effects of Allele Substitutions

In this chapter, we have gone through a number of quantitative genetic models describing the effects of genes on traits. If we had to draw a main, general conclusion in view of those formulae, it would probably be that they model precisely how the variability within a gene may affect the phenotype (as different genotypes may lead to different phenotypes, i.e., different genotypic values). To do so, the models are based on parameters representing the effects of the alleles within genes. To be even more precise, the effects of alleles are implemented in the models as relative effects, indeed, as effects of allele substitutions within a gene. Thus, the models developed in this chapter illustrate that the effect of a gene in quantitative genetics terms is none other than the set of effects of allele substitutions within that gene.

Precisely because the genetic effects are effects of allele substitutions, a one-locus two-allele genetic system needs one only additive effect to be modeled. There is no need of one additive effect for each allele, but one for the pair of alleles—for describing how substituting one allele for the other affects the trait. The dominance effect is also, and actually more obviously, a property of a pair of allels. Following Fisher's (1918) advice, we have already highlighted the genetic effects as effects of allele substitutions in the previous two chapters.

4.11.2 Substitutions in Individuals Versus in Populations

Fisher's (1918) early concern about the divide between the individual and the population levels can also be drawn from the formulae developed in this chapter. We have above resumed (from what has already been explained in the previous two chapters) Fisher's additive/dominance scale, which was originally intended as the simplest yardstick and eventually found to fit a population context—particularly, that of an F_∞ population. We have in particular clarified that a model of genetic effects is not only determined by its genetic effects themselves but also by its reference point and the decomposition of the genetic values that the reference point and the genetic effects jointly lead to.

Indeed, the F_∞ model (i.e., Fisher's additive/dominance scale) provides the same genetic effects as individual-referenced models (which explains why Fisher found it useful to some extent for the individual level) but actually fits, as a whole, a population-referenced model—it describes allele substitutions as performed at an F_∞ population. Therefore, at a time when gene mapping experiments open the door to disclosing the actual effects of genetic variability on traits (which were completely out of reach at the birth of quantitative genetics and for a long time afterwards), it is sensible to completely clarify when we deal with those effects at the Individual and at the population levels.

In order to prevent confusion between the individual and the population levels, we adhere to represent the parameters of the models following an alphabet code. Latin letters, like R, a, and d, shall denote the individual level, whereas Greek letters, like μ, α, and δ, shall denote the population level. Chapter 2 fits to this code mostly—since a and d, coming from Fisher's additive/dominance yardstick, coincide with genetic effects at the individual level. In Chap. 3, however, we have used a mixture of Latin and Greek letters (μ, a, and d), following the most common notation used by the original authors of the models there presented, (see e.g. Zeng et al. 2005). Such mixture reflects, in our view, that the distinction between the individual and the population levels Fisher (1918) was concerned about has remained an issue hitherto.

Additionally, we have shown that the theoretical proposal made in this chapter is consistent with two other concerns raised by Fisher already in his seminal publication (Fisher 1918). On the one hand, our theoretical proposal resolves the uncomfortable polarity Fisher noticed in his original developments, which actually remained in twenty-first century genetic modeling (e.g., Zeng et al. 2005; see Álvarez-Castro and Carlborg 2007). On the other hand, the genetic effects in these models do not only reflect effects of allele substitutions but are in fact effects of allele substitutions at the individual level.

4.11.3 From One Biallelic Locus to Complex Genetic Architectures

We have in general dealt with the different situations separately but have indicated how different increases in complexity can be dealt with at the same time, actually enabling the modeling of innumerable genetic architectures. In fact, we have shown that the models can very easily get to be large even for relatively simple cases, as the model for two loci and five alleles in total (Expressions 4.15 and 4.16). The key tool to combine different factors (whether genetic or environmental) is the Kronecker product of the single-factor genetic-effects design matrices, as introduced in the multiple loci section. In the end, we will not have to handle large expressions at all—we just have to know how the expressions have to be built for programming computers for dealing with them for us.

We have paid special attention to the description of the first case—the basic one-locus two-allele model—which we have used for an in-depth explanation of the meaning of the parameters. From that basis, we have resumed both the distinction between the individual and the population levels and the change-of-reference operation. We have also gone through the imprinting case in some detail, as it is particularly tricky from the modeling perspective—with, strictly speaking, four allelic variants, but not fitting at all to a four-allele model.

The imprinting model was anyway not the only case where a choice can be made about how exactly to develop an individual-referenced model of genetic effects. Also in the sex-linked and sex-interaction cases—which, incidentally, are here published for the first time—it is possible to model some genetic effects as departures from expected values (thus, interactions) or not. We have mentioned that the second choice (modeling explicit interactions) entails an advantage for the development of tests to assess whether the interaction actually occurs—i.e., whether it is significantly different from zero. However, those tests will be done from population-referenced models, which we shall develop in the following chapters—based on the individual-referenced ones developed in this chapter.

4.11.4 Genetic Architectures in Populations

Beyond having to consider the different genetic architectures already dealt with in this chapter, the population-referenced models in the following chapters shall have to consider the different particularities specifically affecting the population level. In particular, developments shall have to be provided for properly dealing with departures from the Hardy-Weinberg proportions at the single-locus level, departures from linkage equilibrium at the multilocus level, and departures from random association between loci and environments at the multiple factors (whether genetic or environmental) level.

The population-referenced models presented in the forthcoming chapters shall also adhere to the matrix notation, thus being straightforwardly adapted to the change-of-reference operation. That was actually the main motivation for the development of the natural and orthogonal interactions (NOIA) model (Álvarez-Castro and Carlborg 2007). Here, we have dealt with the branch of NOIA dealing with the individual level, which was called "natural" because its parameters are effects of allele substitutions performed from real genotypes—thus fitting to an evident biological interpretation. In the following chapters, we shall deal with the population-reference branch of NOIA, which is called "orthogonal" because this is a statistical property such models have to fulfill for their parameters to fit their intended biological interpretation, as well as for the models to enable the development of proper gene-mapping tools.

In the same vein as for the models addressing the individual level, the genetic effects of the models for the population level reflect the effect of a gene (or, in general, of a genetic architecture) by means of effects of allele substitutions. However, those effects are, at the population level, averages of the parameters of the individual level performed over the genetic composition of a population. That concept has been already introduced in Chaps. 2 and 3 and shall be addressed more generally in the following chapters.

References

Álvarez-Castro JM (2014) Dissecting genetic effects with imprinting. Front Ecol Evol 2:51

Álvarez-Castro JM (2016) Genetic architecture. In: Wolf JB (ed) Encyclopedia of evolutionary biology. Oxford Academic Press, Oxford, pp 127–135

Álvarez-Castro JM, Carlborg Ö (2007) A unified model for functional and statistical epistasis and its application in quantitative trait loci analysis. Genetics 176:1151–1167

Álvarez-Castro JM, Yang R-C (2011) Multiallelic models of genetic effects and variance decomposition in non-equilibrium populations. Genetica 139:1119–1134

Cheverud JM, Routman EJ (1995) Epistasis and its contribution to genetic variance components. Genetics 139:1455–1461

Cockerham CC (1954) An extension of the concept of partitioning hereditary variance for analysis of covariances among relatives when epistasis is present. Genetics 39:859–882

Crouch DJM, Bodmer WF (2020) Polygenic inheritance, GWAS, polygenic risk scores, and the search for functional variants. Proc Natl Acad Sci U S A 117:18924–18933

Fisher RA (1918) The correlation between relatives on the supposition of Mendelian inheritance. Trans. Roy. Soc. Edinburgh 52:339–433

Franklin R, Gosling RG (1953) Molecular configuration in sodium thymonucleate. Nature 171: 740–741

Hansen TF, Wagner GP (2001) Modeling genetic architecture: a multilinear theory of gene interaction. Theor Pop Biol 59:61–86

Kawahara M, Wu Q, Takahashi N, Morita S, Yamada K, Ito M, Ferguson-Smith AC, Kono T (2007) High-frequency generation of viable mice from engineered bi-maternal embryos. Nat Biotechnol 25:1045–1050

Kempthorne O (1954) The correlation between relatives in a random mating population. Proc R Soc Lond B Biol Sci 143:102–113

Kono T, Obata Y, Wu Q, Niwa K, Ono Y, Yamamoto Y, Park ES, Seo JS, Ogawa H (2004) Birth of parthenogenetic mice that can develop to adulthood. Nature 428:860–864

Ma J, Xiao F, Xiong M, Andrew AS, Brenner H, Duell EJ, Haugen A, Hoggart C, Hung RJ, Lazarus P, Liu C, Matsuo K, Mayordomo JI, Schwartz AG, Staratschek-Jox A, Wichmann E, Yang P, Amos CI (2012) Natural and orthogonal interaction framework for modeling gene-environment interactions with application to lung cancer. Hum Hered 73:185–194

Mantey C, Brockmann GA, Kalm E, Reinsch N (2005) Mapping and exclusion mapping of genomic imprinting effects in mouse F2 families. J Hered 96:329–338

Morgan TH (1910) Sex limited inheritance in drosophila. Science 32:120–122

Thoday JM (1961) Location of Polygenes. Nature 191:368–370

Tiwari HK, Elston RC (1997) Deriving components of genetic variance for multilocus models. Genet Epidemiol 14:1131–1136

Van Der Veen JH (1959) Tests of non-allelic interaction and linkage for quantitative characters in generations derived from two diploid pure lines. Genetica 30:201–232

Watson JD, Crick FHC (1953) A structure for deoxyribose nucleic acid. Nature 171:737–738

Wolf JB, Cheverud JM, Roseman C, Hager R (2008) Genome-wide analysis reveals a complex pattern of genomic imprinting in mice. PLoS Genet 4:e1000091

Yang R-C, Álvarez-Castro JM (2008) Functional and statistical genetic effects with miltiple alleles. Curr Top Genet 3:49–62

Zeng ZB, Wang T, Zou W (2005) Modeling quantitative trait Loci and interpretation of models. Genetics 169:1711–1725

Genetic Effects in Populations Under Linkage Equilibrium

5

Abstract

In this chapter, models of genetic effects at the population level are provided that cover all the genetic architectures considered in the previous chapter at the individual level—a biallelic locus, multiple loci, multiple alleles, gene-environment interaction, sex-linked loci, gene-sex interaction, and imprinting. With this, the two complementary faces of the natural and orthogonal interactions (NOIA) model are put together. NOIA unifies, under a common mathematical framework, the wide range of parameters needed for in-depth quantitative genetic analysis—the natural and the orthogonal effects of allele substitutions, reflecting properties of the genetic system at, respectively, the individual and the population levels. The concept of orthogonality is here dissected both from the mathematical and the biological perspectives. Some of the results here provided have not been published previously, while some others have been obtained using different approaches in previous publications. In order to facilitate the understanding of this theory, all the results are here developed following a common regression framework. These results open new possibilities of quantitative genetic analyses, particularly for gene-mapping studies. In order to cover a full range of population facts, these models must be further implemented with departures from linkage equilibrium and non-random associations between genes and environments, which has been deferred to the next chapter.

5.1 Introduction

The birth of genetics set in motion an intense debate at the very beginning of the twentieth century, mostly centered on assessing the compatibility of the work of Mendel (which had just been rediscovered) and Darwinian gradual evolution by natural selection (Provine 1971; Stoltzfus and Cable 2014; see the previous chapters of this book). The work of Fisher (1918, 1930) established a sound conceptual and

J. M. Álvarez-Castro, *Genes, Environments and Interactions*,
https://doi.org/10.1007/978-3-031-41159-5_5

mathematical basis to support that compatibility, while in turn sparking another intense debate, this time about the role of epistasis and drift in the evolutionary process. Fisher and Haldane deemed selection to most efficiently act over large populations (with more allele variants with additive effects), whereas Wright found it more plausible that evolution would burst with a combination of epistasis and moderate genetic drift (see e.g. Provine 1971, 1986).

The aforementioned events started at a time when the disclosure of the molecular basis of Mendel's hereditary factors still seemed like a pipe dream. When, later on, Kempthorne (1954) and Cockerham (1954) implemented Fisher's modeling of the genetic effects to analyze the resemblance between relatives with epistasis, the responsible molecule had at last been identified (Avery et al. 1944; Hershey and Chase 1952). However, the understanding of the functioning of the nucleic acids and, thus, exponential growth of molecular genetics was only in its infancy (Franklin and Gosling 1953; Watson and Crick 1953). Therefore, more time would have to pass for Kempthorne (1954) and Cockerham's (1954) theoretical achievements to attain significant applicability in practice.

Several decades later, the advent of quantitative trait locus (QTL) analysis (after the work of Lander and Botstein 1989) showed the need to improve the adequacy of the existing models of genetic effects to the practice of mapping experiments. As the one-locus models seemed possible to grasp, much of the motivation to improve mapping techniques came from updating the models of genetic effects to properly account for epistasis (e.g., Cheverud and Routman 1995; Tiwari and Elston 1997; Hansen and Wagner 2001; Yang 2004; Zeng et al. 2005; Álvarez-Castro and Carlborg 2007).

The aforementioned task required a careful revision of even the simplest models, in order to avoid carrying any inaccuracies that hamper the study of the more complex situations, as well as to set the basis for general theoretical framework implementing all the advances that were initially made separately. Álvarez-Castro and Carlborg (2007) proposed the NOIA model precisely with those objectives in mind. NOIA implements Tiwari and Elston's (1997) matrix notation (revised in Chap. 3) as a suitable framework to formulate the models fitting to both the individual and the population levels.

In what regards the individual level, NOIA takes the baton of Hansen and Wagner's (2001) concepts of allele substitutions performed from an individual genotype and the change-of-reference operation, which in turn were inspired by the concept of physiological epistasis by Cheverud and Routman (1995). The first proposal of NOIA addressed the case of an arbitrary number of biallelic loci, with arbitrary dominance and epistasis (Álvarez-Castro and Carlborg 2007). The expressions shown in the previous chapter constitute the current formulation of NOIA at the individual level.

Concerning now the population level, we have made a revision of the classical models in Chap. 2. As for the models in tune with the gene-mapping era, we have presented Zeng et al.'s (2005) proposal for an arbitrary number of biallelic loci with arbitrary dominance and epistasis under Hardy-Weinberg proportions and linkage equilibrium in Chap. 3. Then, the first proposal of NOIA (Álvarez-Castro and

Carlborg 2007) extended that framework to departures from the Hardy-Weinberg proportions, thus embracing together the conditions considered separately by Zeng et al. (2005) and previously by Yang (2004)—who addressed the two-locus two-allele case with arbitrary dominance and epistasis at populations with arbitrary departures from the Hardy-Weinberg proportions and under linkage equilibrium.

In this way, the first proposal of NOIA (Álvarez-Castro and Carlborg 2007) not only unified and generalized previous models of genetic effects, both at the individual and at the population levels, but above all set up an apposite basis for gradually adding up further implementations, particularly encompassing the context of gene-mapping experiments. As mentioned above, the previous chapter gathers the current implementations of NOIA at the individual level. This chapter and the following one give account of the current implementations of NOIA at the population level. All genetic architectures considered in the previous chapter at the individual level are developed here at the population level. Concerning population facts, arbitrary departures from the Hardy-Weinberg proportions are accounted for in this chapter, while the extensions to arbitrary departures from linkage equilibrium and from random associations between genes and environments are deferred to the next one. Before addressing the models themselves, we shall briefly introduce the method of choice to develop them.

5.2 Regression Framework

In Chap. 2, we have revised the most important facts of the classical models of genetic effects. These models implement parameters with biological meaning of practical use in experimental populations. Although the models are based on a description of the relationship between genotype and phenotype—i.e., on the genotype-to-phenotype (GP) map—accounting also for the effects of environmental factors, they provide applications that only require data on phenotypes and relatedness, i.e., without any direct knowledge about the genetic basis of the trait under study. It is thus understandable that the classical models of genetic effects would in general not need to be very precise about the genetic basis of the trait.

In this chapter and the following one, we shall focus on models appropriate both to work with particular, possibly complex genetic architectures and to actually disclose those genetic architectures in the context of gene-mapping experiments. Therefore, it will be crucial that these models implement accurately diverse and complex genetic architectures. For instance, and as pointed out at the very beginning of this book, we would not like our mould to suffer from limitations potentially imposing preset shapes on the object we intend to bare. Thus our aim in this chapter and the next one is to develop a GP map at the population level along the lines of the classical models of genetic effects presented in Chap. 2 (and, more in particular, of the implementations of them presented in Chap. 3), but accurately embracing much more general situations.

There have been used various (more or less different) approaches to build-up models of genetic effects at the population level. Yang (2004) worked out a two-locus two-allele model with epistasis to be used in gene-mapping experiments, based on statistical contrasts previously designed by Cockerham (1954). Zeng et al. (2005) opted for developing multiple-locus two-allele models by setting one-locus regressions based on different genetic-effect design variables, which could be allele-frequency dependent, thus leading to genetic-effect design matrices appropriate to model populations with different allele frequencies, as explained more in detail below. Álvarez-Castro and Carlborg (2007) and Álvarez-Castro (2014) obtained genotype-frequency dependent models by analyzing in detail the original regression of the genotypic values on the gene-content proposed by Fisher (1918), for implementing departures from the Hardy-Weinberg proportions in multilocus models and imprinting, respectively. In this chapter, we shall follow the approach described by Kempthorne (1957), as used, e.g., by Álvarez-Castro and Yang (2011) for developing a model with multiple alleles and arbitrary departures from the Hardy-Weinberg proportions.

But, what does it exactly mean that a model (or its genetic-effects design matrix) is appropriate for a particular population? For the parameters of a model of genetic effects at the population level to fit to their intended biological meaningm they have to be averages (over the population under study) of the parameters of the corresponding model at the individual level. As already realized by Fisher (1918), this implies that they fit to a particular linear regression framework, as explained in Chap. 2. And this in turn implies—as also explained in Chap. 2—that the genetic effects are obtained as independent, orthogonal estimates. Let us then first take a quick look at the mathematical concept of orthogonality.

5.2.1 When Is a Model Orthogonal at a Population?

As explained, e.g., by Álvarez-Castro and Carlborg (2007), a convenient way to assess whether a model of genetic effects is orthogonal in the context of a given population is to perform a multiplication using its genetic-effects design matrix. In particular, it must be checked whether the outcome of the following matrix multiplication is a diagonal matrix:

$$\mathbf{S}^{\mathrm{T}}\mathbf{P}\mathbf{S}, \tag{5.1}$$

where $\mathbf{S}$ is a genetic-effects design matrix and $\mathbf{P}$ is a diagonal matrix with the frequencies of the population in question at the diagonal, i.e., $\mathbf{P} = diag(p_{ij})$. A few examples follow.

Let us take the very well-known F_2 case, with frequencies $p_{11} = 1/4$, $p_{12} = 1/2$, and $p_{22} = 1/4$, of genotypes A_1A_1, A_1A_2, and A_2A_2, respectively. Applying Expression 5.1 to the genetic-effects design matrix of the F_2 case Expression 3.5, we obtain:

$$\mathbf{S}_{F2}^{T}\mathbf{P}_{F2}\mathbf{S}_{F2}=\begin{pmatrix}1 & 1 & 1\\ 1 & 0 & -1\\ -½ & ½ & -½\end{pmatrix}\begin{pmatrix}¼ & 0 & 0\\ 0 & ½ & 0\\ 0 & 0 & ¼\end{pmatrix}\begin{pmatrix}1 & 1 & -½\\ 1 & 0 & ½\\ 1 & -1 & -½\end{pmatrix}$$
$$=\begin{pmatrix}1 & 0 & 0\\ 0 & ½ & 0\\ 0 & 0 & ¼\end{pmatrix}, \tag{5.2}$$

which is in fact a diagonal matrix. Note that if we apply this expression to a different population (with different frequencies) then orthogonality does not hold. Let us, for instance, consider a population HW4, under Hardy-Weinberg proportions with the frequency of the first allele being $p_1 = f(A_1) = 0.4$—and thus $p_{11} = 0.16, p_{12} = 0.48$, and $p_{22} = 0.36$. In this case, Expression 5.1 leads to:

$$\mathbf{S}_{F2}^{T}\mathbf{P}_{HW4}\mathbf{S}_{F2}$$
$$=\begin{pmatrix}1 & 1 & 1\\ 1 & 0 & -1\\ -½ & ½ & -½\end{pmatrix}\begin{pmatrix}0.16 & 0 & 0\\ 0 & 0.48 & 0\\ 0 & 0 & 0.36\end{pmatrix}\begin{pmatrix}1 & 1 & -½\\ 1 & 0 & ½\\ 1 & -1 & -½\end{pmatrix}$$
$$=\begin{pmatrix}1 & -0.2 & -0.2\\ -0.2 & 0.52 & 0.1\\ -0.2 & -0.1 & 0.25\end{pmatrix}, \tag{5.3}$$

which no longer is a diagonal matrix.

As explained in Chap. 3, it is Zeng et al.'s (2005) G2A model that can handle this kind of situations. Indeed, if we apply the genetic-effects design matrix of the G2A model Expression 3.7 to $p_1 = 0.4$ (and, thus, $p_2 = 0.6$) and then compute Expression (5.1) with the corresponding Hardy-Weinberg proportions $p_{11} = 0.16$, $p_{12} = 0.48$, and $p_{22} = 0.36$, we obtain:

$$\mathbf{S}_{HW4}^{T}\mathbf{P}_{HW4}\mathbf{S}_{HW4}$$
$$=\begin{pmatrix}1 & 1 & 1\\ 1.2 & 0.2 & -0.8\\ -0.72 & 0.48 & -0.32\end{pmatrix}\begin{pmatrix}0.16 & 0 & 0\\ 0 & 0.48 & 0\\ 0 & 0 & 0.36\end{pmatrix}\begin{pmatrix}1 & 1.2 & -0.72\\ 1 & 0.2 & 0.48\\ 1 & -0.8 & -0.32\end{pmatrix}$$
$$=\begin{pmatrix}1 & 0 & 0\\ 0 & 0.48 & 0\\ 0 & 0 & 0.2304\end{pmatrix}, \tag{5.4}$$

which brings us back to a diagonal matrix.

The applicability of Zeng et al.'s (2005) G2A model is however not completely general and, in particular, constrained to Hardy-Weinberg proportions. In order to illustrate this constraint, let us consider a population with the same allele frequencies

($p_1 = 0.4$ and $p_2 = 0.6$) and an inbreeding coefficient F $= 1.1\bar{6}$, leading to the inbreeding genotype frequencies $p_{11} = 0.2, p_{12} = 0.4$, and $p_{22} = 0.4$. Then, applying Expression 5.1, we obtain:

$$\mathbf{S}_{\mathrm{HW4}}^{\mathrm{T}}\mathbf{P}_{\mathrm{inb}}\mathbf{S}_{\mathrm{HW4}} = \begin{pmatrix} 1 & 1 & 1 \\ 1.2 & 0.2 & -0.8 \\ -0.72 & 0.48 & -0.32 \end{pmatrix}\begin{pmatrix} 0.2 & 0 & 0 \\ 0 & 0.4 & 0 \\ 0 & 0 & 0.4 \end{pmatrix}\begin{pmatrix} 1 & 1.2 & -0.72 \\ 1 & 0.2 & 0.48 \\ 1 & -0.8 & -0.32 \end{pmatrix} = \begin{pmatrix} 1 & 0 & -0.08 \\ 0 & 0.56 & -0.032 \\ -0.08 & -0.032 & 0.2368 \end{pmatrix}, \tag{5.5}$$

which again is not a diagonal matrix. Thus, at a population with the above genotype frequencies, the biological meaning of the parameters of the models is not the desired one (the one explained in Chap. 2).

In the next section, we shall confirm that NOIA can maintain orthogonality with those genotype frequencies and, in general, under arbitrary departures from the Hardy-Weinberg proportions. Before doing so, we hereafter open a subsection to introduce the regression framework based on which all the models of genetic effects at the population level presented in this and the following chapter can be built—as conveniently featured by Kempthorne (1957).

5.2.2 Kempthorne's Regression Framework

As addressed in Chap. 2, Kempthorne (1968) openly criticized the way Fisher (1918) presented his developments explaining resemblance between relatives in terms of Mendelian inheritance. As explained in Chap. 3, he was not alone in getting in trouble when trying to understand Fisher (1918) in detail. In any case, as we in turn find that Kempthorne's indications (e.g., Kempthorne 1957) may still leave some room for further clarification, we hereafter review in detail the key facts of his regression framework (modified from Álvarez-Castro and Crujeiras 2019).

Kempthorne (1957) considered the regression framework for the simple one-locus two-allele case by expressing the genotypic values in terms of the (additive and dominance) genetic effects case as $\mathbf{G} = \mathbf{1}\,\mu + \mathbf{N}\boldsymbol{\alpha} + \boldsymbol{\delta}$, expanding to:

$$\begin{pmatrix} \mathrm{G}_{11} \\ G_{12} \\ G_{22} \end{pmatrix} = \begin{pmatrix} 1 \\ 1 \\ 1 \end{pmatrix}\mu + \begin{pmatrix} 2 & 0 \\ 1 & 1 \\ 0 & 2 \end{pmatrix}\begin{pmatrix} \alpha_1 \\ \alpha_2 \end{pmatrix} + \begin{pmatrix} \delta_{11} \\ \delta_{12} \\ \delta_{22} \end{pmatrix}. \tag{5.6}$$

We recall from the previous chapters that μ is the population mean. Also, from Expression 2.8, we recall that α_1 and α_2 are the average effects of allele

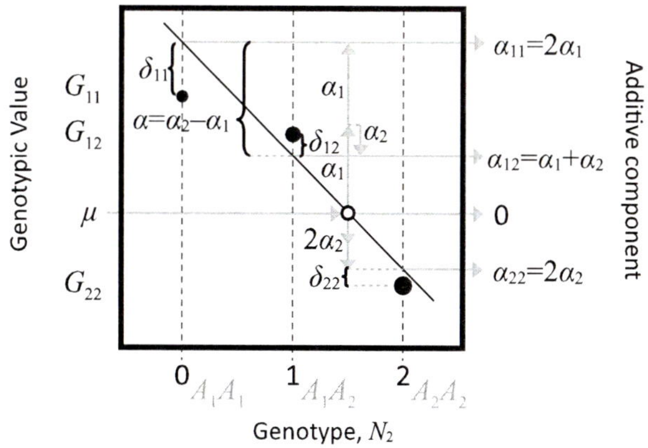

Fig. 5.1 Graphical interpretation of Fisher's (1918) additive and dominance population parameters. The additive parameters are the average effects of allele substitutions, α_i, $i = 1$, 2, and breeding values, α_{ij}, $ij = 11$, 12, 22. The dominance parameters are δ_{ij}, $ij = 11$, 12, 22. The genotypes are represented in the horizontal axis of the graph by their content of allele A_2. The black solid line is the regression (linear approximation) to the genotypic values. This graph is a combination of Figs. 2.2 and 2.3

substitutions—from which the average effect of the gene at the population and breeding values can be obtained as (Expression 2.9) $\alpha = \alpha_2 - \alpha_1$, $\alpha_{ij} = \alpha_i + \alpha_j$, $ij = 11, 12, 22$, respectively and δ_{ij}, $ij = 11, 12, 22$, are the dominance deviations (see Expressions 2.12, 2.13). The graphical interpretation of these parameters is given in Fig. 5.1.

Expression 5.6 indeed sets up a regression framework in which the genotypic values are the dependent variables, the population mean and the average effects of allele substitutions are the explanatory variables, and the dominance deviations are the error terms. Matrices **1** and **N** account for how the explanatory variables relate to the dependent variables. Matrix **1** sets the reference point at the population mean, and matrix **N** just indicates the alleles of each genotype.

For the estimates obtained to achieve their intended biological meaning, regression model in Expression 5.6 has to be solved by first clearing away the term with the mean phenotype using the mean-corrected genotypic values as:

$$\overline{\mathbf{G}} = \mathbf{G} - \mathbf{1}\,\mu, \tag{5.7}$$

and then solving the resulting regression $\overline{\mathbf{G}}=\mathbf{N}\boldsymbol{\alpha} + \boldsymbol{\delta}$. The weighted least-squares (WLS) solution and error terms for this regression may be obtained from its normal equations, $\mathbf{N}^T\mathbf{P}\mathbf{N}\boldsymbol{\alpha}=\mathbf{N}^T\mathbf{P}\overline{\mathbf{G}}$, where **N** is called the design matrix of the regression, $\mathbf{N}^T$ stands for its transpose, and **P**—which as mentioned above contains the population genotypic frequencies ($f(A_iA_j) = p_{ij}$, $j = 1{,}2$, $i \leq j$) in its diagonal, i.e., $\mathbf{P} = diag(p_{ij})$—is called the weight matrix (for a revision of the theory of matrix algebra here applied see, e.g., Harville 1997; Draper and Smith 1998). This way we obtain:

$$\boldsymbol{\alpha} = \tilde{\mathbf{H}}\overline{\mathbf{G}}, \qquad \text{with } \tilde{\mathbf{H}} = \left(\mathbf{N}^{\mathrm{T}}\mathbf{P}\mathbf{N}\right)^{-1}\mathbf{N}^{\mathrm{T}}\mathbf{P}. \tag{5.8}$$

Matrix $\mathbf{H} = \mathbf{N}\tilde{\mathbf{H}}$ is often called the hat matrix of the regression and $\mathbf{M} = \mathbf{I} - \mathbf{H}$ ($\mathbf{I}$ being the identity matrix with the proper dimension, 3×3) is the annihilation matrix. Using the latest matrix the error terms may be equated as:

$$\boldsymbol{\delta} = \mathbf{M}\overline{\mathbf{G}}. \tag{5.9}$$

It is worthwhile to emphasize that Expression 5.7—i.e., the mean-corrected genotypic values—can also be obtained as the WLS solution of a linear regression framework. Such framework is indeed the mean-related part of Expression 5.6, $\mathbf{G} = \mathbf{1}\,\mu + \boldsymbol{\eta}$. Applying the aforementioned steps to this expression, we easily get that vector $\mathbf{1} = (1, 1, 1)^{\mathrm{T}}$ is the design matrix; hence $\mathbf{1}\mathbf{1}^{\mathrm{T}}\mathbf{P}$ is the hat matrix, $\mu = \mathbf{1}^{\mathrm{T}}\mathbf{P}\mathbf{G}$ is the explanatory variable, $\mathbf{I} - \mathbf{1}\mathbf{1}^{\mathrm{T}}\mathbf{P}$ is the annihilation matrix, and $\boldsymbol{\eta} = (\mathbf{I} - \mathbf{1}\mathbf{1}^{\mathrm{T}}\mathbf{P})\mathbf{G}$ is the error term. Therefore, the mean-corrected vector of genotypic values $\overline{\mathbf{G}} = \mathbf{G} - \mathbf{1}\,\mu$ in Expression 5.7 can actually be interpreted as the error term of regression $\mathbf{G} = \mathbf{1}\,\mu + \boldsymbol{\eta}$, since following Expression 5.8 the WLS solution of that regression provides the error term as $\boldsymbol{\eta} = (\mathbf{I} - \mathbf{1}\mathbf{1}^{\mathrm{T}}\mathbf{P})\mathbf{G} = \mathbf{G} - \mathbf{1}\,\mu$.

Thus, regression model in Expression 5.6 leads to express each of the dependent variables (the genotypic values) in terms of the explanatory variables (the population mean and the average effects of allele substitutions) and of the error terms (the dominance deviations) as $G_{ij} = \mu + \alpha_i + \alpha_j + \delta_{ij}$, $j = 1{,}2$, $i \leq j$. It is nevertheless crucial to keep in mind that, for obtaining these biologically meaningful parameters, the solution achieved through two sequential regression steps is hierarchically applied to the regression model in Expression 5.6. The first step is a rather trivial one, $\mathbf{G} = \mathbf{1}\,\mu + \boldsymbol{\eta}$, evidently leading to the error term $\boldsymbol{\eta} = \overline{\mathbf{G}}$. The second one takes the error term of the first step as dependent variable, as $\boldsymbol{\eta} = \overline{\mathbf{G}} = \mathbf{N}\boldsymbol{\alpha} + \boldsymbol{\delta}$. The sequential procedure guarantees the orthogonality of the different layers of parameters (population mean, additive effects, and dominance effects), which is necessary for the parameters to attain their intended biological meaning.

Incidentally, we could feel tempted to consider a more direct, compact way of expressing the relationship entailed in Expression 5.6 as $\mathbf{G} = (\mathbf{1}|\mathbf{N})(\mu|\boldsymbol{\alpha})^{\mathrm{T}} + \boldsymbol{\delta}$, expanding to:

$$\begin{pmatrix} G_{11} \\ G_{12} \\ G_{22} \end{pmatrix} = \left(\begin{array}{c|cc} 1 & 2 & 0 \\ 1 & 1 & 1 \\ 1 & 0 & 2 \end{array}\right) \left(\begin{array}{c} \mu \\ \hline \alpha_1 \\ \alpha_2 \end{array}\right) + \begin{pmatrix} \delta_{11} \\ \delta_{12} \\ \delta_{22} \end{pmatrix}. \tag{5.10}$$

However, it will be crucial to bear in mind that regression model in Expression 5.6, as solved above in Expressions 5.7–5.9, is not equivalent to this new regression model (Expression 5.10). In fact, following the steps above (Expressions 5.8 and 5.9 and related text) the WLS solution of regression in Expression 5.10 can easily be found to be different from the results obtained within Expressions 5.7–5.9, to the point of μ not necessarily being the population mean phenotype any longer. This is to say, we really need to solve Expression 5.6 sequentially in order to obtain the desired estimates, and not in a single step—i.e., not from Expression 5.10. This issue will become even more vital as we increase the complexity of the models, which may lead to additional layers of parameters and, thus, to higher numbers of sequential steps to be applied for achieving the solutions of the corresponding regression models (as it is the case below in this chapter and in the next chapter).

5.3 Multiple Biallelic Loci

5.3.1 One Locus

Let us begin with the one-locus case first, with alleles A_1 and A_2. We have just addressed Kempthorne's (1957) regression framework for a one-locus two-allele case above (Expression 5.6 and related text). Expressions 5.7–5.9 provide the solution to that case, i.e., the parameters of the model of genetic effects at the population level. For building up the one-locus two-allele formulation of NOIA at the population level, we only have to obtain a genetic-effects design matrix out of those results.

For doing so, we shall actually start by providing the inverse of that matrix and, thus, an expression of the kind $\mathbf{E} = \mathbf{S}^{-1}\mathbf{G}$, with $\mathbf{E} = (\mu, \alpha, \delta)^{\mathrm{T}}$ and $\mathbf{G} = (G_{11}, G_{12}, G_{22})^{\mathrm{T}}$. Therefore, each row of the inverse matrix $\mathbf{S}^{-1}$ is filled with the expressions that provide each of the components of $\mathbf{E}$ as a linear combination of the genotypic values, G_{11}, G_{12}, and G_{22}. Thus, the first row of matrix $\mathbf{S}^{-1}$ is trivially (p_{11}, p_{12}, p_{22}), since $\mu = p_{11}G_{11} + p_{12}G_{12} + p_{22}G_{22}$. As recalled above, in virtue of Expression 2.9, the additive genetic effect, α, can be obtained from the average effects of allele substitutions as $\alpha = \alpha_2 - \alpha_1$. The expressions for the average effects of allele substitutions in terms of the genotypic values can be obtained with Expression 5.8, thus providing the second row of matrix $\mathbf{S}^{-1}$. The last row of this matrix always coincides with that of the corresponding matrix at the individual level (see, e.g., Expressions 3.2, 3.6, 3.8), which in this case is (−½, 1, −½). With this, the $\mathbf{S}^{-1}$ matrix and actually the whole expression $\mathbf{E} = \mathbf{S}^{-1}\mathbf{G}$ at the population level can be expressed as:

$$\begin{pmatrix} \mu \\ \alpha \\ \delta \end{pmatrix} = \begin{pmatrix} p_{11} & p_{12} & p_{22} \\ -\dfrac{p_{11}p_2}{2p_1p_2 - \frac{1}{2}p_{12}} & \dfrac{p_{12}(p_1 - p_2)}{4p_1p_2 - p_{12}} & \dfrac{p_1p_{22}}{2p_1p_2 - \frac{1}{2}p_{12}} \\ -\frac{1}{2} & 1 & -\frac{1}{2} \end{pmatrix} \begin{pmatrix} G_{11} \\ G_{12} \\ G_{22} \end{pmatrix}. \tag{5.11}$$

Since $(\mathbf{S}^{-1})^{-1} = \mathbf{S}$, from this expression, the equivalent expression in the $\mathbf{G} = \mathbf{SE}$ form can be obtained, by just inverting matrix $\mathbf{S}^{-1}$, as:

$$\begin{pmatrix} G_{11} \\ G_{12} \\ G_{22} \end{pmatrix} = \begin{pmatrix} 1 & -2p_2 & -\dfrac{p_{12}p_{22}}{2p_1p_2 - \frac{1}{2}p_{12}} \\ 1 & p_1 - p_2 & \dfrac{p_{11}p_{22}}{p_1p_2 - \frac{1}{4}p_{12}} \\ 1 & 2p_1 & -\dfrac{p_{11}p_{12}}{2p_1p_2 - \frac{1}{2}p_{12}} \end{pmatrix} \begin{pmatrix} \mu \\ \alpha \\ \delta \end{pmatrix}. \tag{5.12}$$

These Expressions 5.11 and 5.12 of NOIA have been expressed in different equivalent ways in different publications (e.g., cf. Álvarez-Castro and Carlborg 2007; Álvarez-Castro 2016). Indeed, the denominator $2p_1p_2 - \frac{1}{2}p_{12}$ can be expressed just as well as $p_{11}p_2 + p_{22}p_1$. In any case, as we have claimed that NOIA generalizes the G2A model, from Expressions 5.11 and 5.12, it should be possible to obtain the G2A model (Expressions 2.7, 2.8), which, in turn, embraces the F_2 model (Expressions 2.5, 2.6), as a particular case. There is, however, a slight difference between the result of constraining NOIA to Hardy-Weinberg proportions and the G2A model, as discussed by Álvarez-Castro and Carlborg (2007). We explain that difference hereafter.

By just replacing p_{ij} by p_ip_j in Expression 5.12, NOIA gets restricted to Hardy-Weinberg proportions. After some basic algebra, it is possible to show that NOIA restricted to Hardy-Weinberg provides the same genetic-effects design matrix as that of Zeng et al.'s (2005) G2A model (Expression 3.8), but for a reverse in the signs of the scalars at the second column of the matrices. This reverse in sign is analogous to the one discussed in the previous chapter between the NOIA model at the individual level and the F_∞ model. Thus, the G2A model adheres to the conventional assumption that allele A_2 always performs worse than allele A_1 (see, e.g., Falconer and MacKay 1996). Thus as Álvarez-Castro and Carlborg (2007) point out, Zeng et al.'s (2005) "estimates reflect the absolute value of the additive effects, by reporting the positive average decrement of an allele substitution from allele [A_1] to allele [A_2]."

As discussed in the previous chapter, that is not a consistent formulation, and it is particularly inconvenient when considering epistasis. As it happens, the G2A model has been developed with particular emphasis on the analysis of epistasis (see Zeng et al. 2005). The point is that epistasis may reverse the direction of the effects of allele substitutions (i.e., their "polarity", as referred to in previous chapters)—a mechanism known as sign epistasis (Weinreich et al. 2005). When this happens, the additive effects of allele substitutions of the G2A model are positive when they decrease the phenotype and become negative when epistasis reverse them so that they increase the phenotype. Thanks to the sign change, the additive effects of the NOIA model at the population level simply reflect the direction of their effects on the phenotype (they are positive when they increase the phenotype and negative when they decrease it), just as the NOIA model at the individual level, as discussed in the previous chapter.

Thus, NOIA provides the G2A model as a particular case, but for a convenient sign correction. The same applies to other commonly used models like the F_2, the F_∞, and Cheverud and Routman's (1995) UWR models (Expressions 3.5, 3.1, 3.10, respectively). All these models can be obtained from NOIA by just applying the population frequencies of each of those models to Expression 5.12, as made explicit above for the G2A model.

5.3.2 Orthogonal Decomposition of the Genotypic Variance

In the first place, let us now retake the case with departures from the Hardy-Weinberg proportions considered in Expression 5.5 that the G2A model could not handle orthogonally. By applying instead the genetic-effects design matrix of NOIA (Expression 5.12) with the inbreeding frequencies of that case (and so indicated in the subscripts), we obtain:

$$\begin{aligned}
&\mathbf{S}_{\text{inb}}^{\text{T}}\mathbf{P}_{\text{inb}}\mathbf{S}_{\text{inb}} \\
&= \begin{pmatrix} 1 & 1 & 1 \\ -1.2 & -0.2 & 0.8 \\ -0.5714 & 0.5714 & -0.2857 \end{pmatrix} \begin{pmatrix} 0.2 & 0 & 0 \\ 0 & 0.4 & 0 \\ 0 & 0 & 0.4 \end{pmatrix} \begin{pmatrix} 1 & -1.2 & -0.5714 \\ 1 & -0.2 & 0.5714 \\ 1 & 0.8 & -0.2857 \end{pmatrix} \\
&= \begin{pmatrix} 1 & 0 & 0 \\ 0 & 0.56 & 0 \\ 0 & 0 & 0.2286 \end{pmatrix},
\end{aligned} \tag{5.13}$$

which is a diagonal matrix, thus showing that the first proposal of NOIA (Álvarez-Castro and Carlborg 2007, shown in Expressions 5.11, 5.12) is indeed orthogonal in the face of departures from the Hardy-Weinberg proportions.

Next, note that the estimates of the explanatory variables of regression in Expression 5.6, as obtained in Expressions 5.7–5.9, can be easily retrieved from Expression 5.12. Indeed, this expression provides the decomposition of the genotypic values (and, hence, also of the genetic variance) into additive and dominance parts, $G_{ij} = \mu + \alpha_{ij} + \delta_{ij}$, $j = 1,2$, $i \leq j$ (Expression 2.22), from which the average effects of allele substitutions can be deduced from the breeding values just as $\alpha_i = \frac{1}{2}\alpha_{ii}$, $i = 1,2$—although, to be precise, the additive part of the decomposition, α_{ij}, can be called breeding values only as long as the Hardy-Weinberg proportions are met.

Finally, by just computing the variances of the different components, the above decomposition of the genotypic values leads to an orthogonal decomposition of the genotypic variance into the additive and the dominance variances, $V_G = V_A + V_D$ (Expression 2.23). Because the model is orthogonal, the resulting components (particularly, the additive and dominance components) are independent and, thus, their covariance is nil: $cov(A,D) = 0$. Otherwise, this covariance would have to be taken into account when computing a (non-orthogonal) decomposition of the genotypic variance since, in the general case, $V_G = V_A + V_D + cov(A,D)$.

5.3.3 Multiple Loci

One advantage of developing a one-locus model of genetic effects in matrix notation (sensu Tiwari and Elston 1997) is that it enables a straightforward extension to multiple loci with epistasis, as shown in Chap. 3 (Expressions (3.3, 3.4) and related text). Thus, Expressions 5.11 and 5.12 implicitly entail the case of arbitrary biallelic loci with arbitrary departures from the Hardy-Weinberg proportions and arbitrary epistasis. The only difference with what was done in Chap. 3 is that, if we want to keep the model flexible in what regards the population frequencies, we now have to implement the notation to indicate to which locus each frequency refers. That can be done, for instance, using superscripts so that p_{ij}^{l} denotes the frequency of genotype $A_i^l A_j^l$ of locus A^l.

The Kronecker product of orthogonal matrices also is an orthogonal matrix (e.g., Van Loan 2000). For this reason, the Kronecker product used in the previous chapter with models at the individual level remains valid in this chapter with models at the population level. This property has been used to extend to multiple loci both the G2A model and NOIA. More to the point, the first proposal of NOIA (Álvarez-Castro and Carlborg 2007) generalized not only Zeng et al's (2005) G2A model, for arbitrary biallelic, putatively epistasic loci under the Hardy-Weinberg proportions but also Yang's (2004) model, for two biallelic loci with putative departures from the Hardy-Weinberg proportions. As mentioned above, we shall address the extension to account for arbitrary departures from linkage equilibrium frequencies in the next chapter.

5.4 Multiple Alleles

The extension of Kempthone's (1957) regression framework (Expression 5.6) to multiple alleles can be done by just enlarging the dimension of the vectors and matrices as required by the number of alleles to be considered and keeping the logic of the design matrix $\mathbf{N}$ (the additive part of the genotypic value G_{ij} is $\alpha_{ij} = \alpha_i + \alpha_j$) when enlarging it (Yang and Álvarez-Castro 2008). For instance, with three alleles, we would have the same general expression for the regression framework as for two alleles, $\mathbf{G} = \mathbf{1}\,\mu + \mathbf{N}\boldsymbol{\alpha} + \boldsymbol{\delta}$, but now expanding to:

$$\begin{pmatrix} G_{11} \\ G_{12} \\ G_{22} \\ G_{13} \\ G_{23} \\ G_{33} \end{pmatrix} = \begin{pmatrix} 1 \\ 1 \\ 1 \\ 1 \\ 1 \\ 1 \end{pmatrix} \mu + \begin{pmatrix} 2 & 0 & 0 \\ 1 & 1 & 0 \\ 0 & 2 & 0 \\ 1 & 0 & 1 \\ 0 & 1 & 1 \\ 0 & 0 & 2 \end{pmatrix} \begin{pmatrix} \alpha_1 \\ \alpha_2 \\ \alpha_3 \end{pmatrix} + \begin{pmatrix} \delta_{11} \\ \delta_{12} \\ \delta_{22} \\ \delta_{13} \\ \delta_{23} \\ \delta_{33} \end{pmatrix}. \tag{5.14}$$

Expressions 5.7–5.9 and related text keep on providing the solution to the regression, by just taking into account that, with more alleles, the weight matrix,

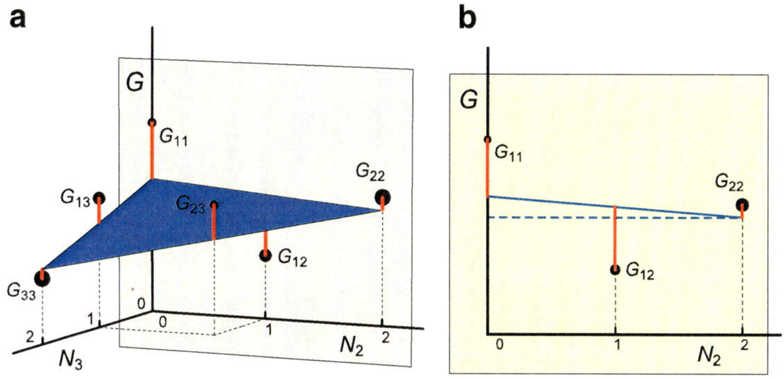

Fig. 5.2 Graphical interpretation of the formulation (at the population level) of the one-locus three-allele genetic system in Table 5.1. (**a**) The genotypic values (G_{ij}) are spheres whose size represents the corresponding genotypic frequencies (p_{ij}). The blue surface is the linear regression of the genotypic values on the number of alleles A^2 and A^3 (N_2 and N_3, respectively). The red vertical lines stand for the deviations of the genotypic values and the breeding values, at the regression plane. The slice corresponding to the axes for the genotypic values and the content of allele A^2 (shadowed area) is detached in Fig. 5.2b. (**b**) Slice corresponding to the axes G and N_2 (shadowed area in Fig. 5.1a). The solid line marks the intersection with the regression plane (blue surface in Fig. 5.1a). The dashed blue line is the corresponding linear regression in the absence of allele A^3—the remaining genotypes keeping their relative frequencies (from Álvarez-Castro et al. 2012)

P $= diag(p_{ij})$ contains the population genotypic frequencies of more genotypes (in particular, with k alleles, $p_{ij} = f(A_iA_j), j = 1, 2, \ldots, k, i \leq j$).

The solution to regression framework in Expression 5.14 enables the construction an inverse genetic-effects design matrix, $\mathbf{S}^{-1}$, with multiple alleles, at the population level. Since the definitions of the population mean and of the dominance genetic effects in terms of the genotypic values are straightforward ($\mu = \Sigma p_{ij}G_{ij}$ and $\delta^{ij} = G_{ij} - ((G_{ii} + G_{jj})/2), j = 1,2,..,k, i \leq j$, respectively), we only have to obtain the remaining additive genetic effects (accounting from effects of allele substitutions between pairs of alleles) from the solution of regression framework in Expression 5.14—particularly, from Expression 5.8—as $\alpha^{ij} = \alpha_j - \alpha_i$.

It is at this point worthwhile to recall that the use of superscripts to denote these additive and dominance genetic effects, α^{ij} and δ^{ij}, respectively, is necessary to prevent confusion with the additive and dominance portions of the decomposition of the genotypic value G_{ij}, $\alpha_{ij} = \alpha_i + \alpha_j$ and δ_{ij}, respectively. The meaning of these parameters was much easier to track in the models of genetic effects at the individual level (Chap. 4), whose parameters reflect effects of allele substitutions from an individual genotype (see, e.g., Fig. 4.2). At the population level, the genetic effects, α^{ij} and δ^{ij}, are conceptually more abstract as they are averages over the population of the ones at the individual level, a^{ij} and d^{ij}, but they are consequential as they reveal crucial properties of the populations. For instance, the additive effect of a couple of alleles A_i and A_j, α^{ij}, reflects how much the relative frequencies of those alleles may be altered by selection on the phenotype (as already pointed out in the text related to Expression 2.10, although no superscripts are needed there, with one only possible pair of different alleles to be considered).

Table 5.1 Genotype values, genotype frequencies, and additive components for the six genotypes of the one-locus three-allele case illustrated in Fig. 5.1 (from Álvarez-Castro et al. 2012)

	11	12	22	13	23	33
G_{ij}	1.5	0.5	1	1	1	0.5
p_{ij}	0.05	0.15	0.30	0.15	0.05	0.30
α_{ij}	0.26	0.18	0.10	0.01	−0.07	−0.23

Figure 5.2 and Table 5.1 (both from Álvarez-Castro et al. 2012) illustrate a three-allele case at the population level. The genetic effects at the population level with multiple alleles essentially hold the meaning of the two-allele case (Figs. 2.2 and 2.3). Figure 5.2 particularly shows that, however, the genetic effect of a two-allele case (blue dashed line) does not necessarily coincide with that those two alleles would have when more alleles are added (blue solid line), even when the relative frequencies of the original two alleles are kept. In other words, at the population level, the properties of a pair of alleles cannot be generally isolated from what happens to rest of the alleles.

Multiallelic loci can be implemented within multilocus systems just as explained in the previous section for biallelic loci. The Kroneker product of genetic-effects design matrices (or of inverses of them) enables extensions to multiple loci with arbitrary epistasis regardless of the numbers of alleles of each of the loci. Such extensions assume linkage equilibrium—i.e., provide orthogonal multilocus models only under strict linkage equilibrium. The case of departures from linkage equilibrium is dealt with in the next chapter. As explained in the previous chapter (Expressions 4.14–4.16 and related text in Chap. 4), this may lead to models with high numbers of parameters. Also in the next chapter, we shall comment on cases where the number of parameters needs to be cut down.

5.5 Gene-Environment Interaction

It is possible to develop a regression framework along the lines of Expression 5.6 for an environmental factor alone. For an environmental factor E with environmental alternatives E_1 and E_2, with effects ε_1 and ε_2, respectively, such framework could be set up just as $\mathbf{G} = \mathbf{1}\,\mu + \boldsymbol{\varepsilon}$, expanding to:

$$\begin{pmatrix} G_1 \\ G_2 \end{pmatrix} = \begin{pmatrix} 1 \\ 1 \end{pmatrix}\mu + \begin{pmatrix} \varepsilon_1 \\ \varepsilon_2 \end{pmatrix}. \tag{5.15}$$

Although we keep the notation of the dependent variables in vector $\mathbf{G}$, they now stand for environmental values—instead of genotypic values.

As in the equivalent model at the individual level (see the section on gene-environment interaction in Chap. 4), we are initially considering one only genotype, G, with different genotypic values under different alternatives for an environmental variable. Those are the dependent variables, gathered in vector **G**. The explanatory variable is just the population mean and the environmental alternatives (analogous to alternative alleles), gathered in the vector **ε**, are the error terms. As such, regression framework in Expression 5.15 is however trivial since it just implies detaching the mean-corrected environmental values (using Expression 5.7, just leading to $\varepsilon_i = \bar{G}_i = G_i - \mu$, $i = 1, 2$).

Although not tremendously revealing, the solution of the above regression framework Expression 5.15 is useful for us as a reassuring intermediate step to build up the environmental-effects design matrix via its inverse, thus following the standard procedure to obtain the genetic-effects design matrices in the previous sections. In the current case, with the effects of the environmental alternatives, ε_i, being equal to the mean-corrected environmental values, $\bar{G}_i$, and defining the environmental effect (the effect of changing from one environmental alternative to another) as $\varepsilon = \varepsilon_2 - \varepsilon_1$, it is trivial to derive ε in terms of the scalars of **G** as $\varepsilon = \bar{G}_2 - \bar{G}_1 = G_2 - G_1$.

It is at this point interesting to note that, as in the previous sections, the upper-level effect is defined, in the model at the population level, in the same way as in the corresponding model at the individual level (Expression 4.18 and related text in Chap. 4). The only difference here is that the upper-level effect is virtually the only effect—not considering the population mean as an effect, strictly speaking, although it is the first scalar of the vector of (commonly genetic, here environmental) effects, **E**. Incidentally, also note that this capital letter "E" does not stand for "environmental," but for "effects," as we also use **E** to denote the vector of genetic effects in the previous sections and in the previous chapters.

Herewith, the model of environmental effects can be expressed in terms of the inverse of the environmental-effects design matrix as $\mathbf{E} = \mathbf{S}^{-1}\mathbf{G}$, expanding to:

$$\begin{pmatrix} \mu \\ \varepsilon \end{pmatrix} = \begin{pmatrix} q_1 & q_2 \\ -1 & 1 \end{pmatrix} \begin{pmatrix} G_1 \\ G_2 \end{pmatrix}, \tag{5.16}$$

where q_i, $i = 1, 2$, are the population frequencies of the individuals at environments E_i, $i = 1, 2$. Regression ni Expression 5.15 has just provided us with the first row on the matrix in Expression 5.16, the second row being equal to the equivalent model at the individual level Expression 4.18.

Then, by inverting $\mathbf{S}^{-1}$, the equivalent expression in terms of the environmental-effects design matrix, **S**, can easily be obtained as $\mathbf{G} = \mathbf{SE}$, expanding to:

$$\begin{pmatrix} G_1 \\ G_2 \end{pmatrix} = \begin{pmatrix} 1 & -q_2 \\ 1 & q_1 \end{pmatrix} \begin{pmatrix} \mu \\ \varepsilon \end{pmatrix}. \tag{5.17}$$

As explained when dealing with the equivalent models at the individual level (Expressions 4.17 and 4.18), Expressions 5.16 and 5.17 also fit a haploid, biallelic locus but, when applied to an environmental variable, for them to become really meaningful, a genetic context must be added up, i.e., they have to be combined with the genetic basis of the trait and, thus, to make them account for potential gene-environment interactions. Also as for the models at the individual level, this can be done by just computing the Kronecker product of the environmental-effects design matrix with the genetic-effects design matrix.

As an illustrative example, let us consider a simple one-locus two-allele model (Expression 5.12) with the environmental effects considered right above (Expression 5.17). The combined genetic and environmental effects can thus be described using the Kronecker product (see Expressions 3.3 and 3.4 and related text in Chap. 3 and Expressions 4.21 and 4.22 in Chap. 4). The genetic/environmental-effects design matrix is thus computed as the Kronecker product of the environmental-effects design matrix (Expression 5.17) and the genetic-effects design matrix (Expression 5.12) as:

$$\mathbf{S} = \begin{pmatrix} 1 & -q_2 \\ 1 & q_1 \end{pmatrix} \otimes \begin{pmatrix} 1 & -2p_2 & -\dfrac{p_{12}p_{22}}{2p_1p_2 - \frac{1}{2}p_{12}} \\ 1 & p_1 - p_2 & \dfrac{p_{11}p_{22}}{p_1p_2 - \frac{1}{4}p_{12}} \\ 1 & 2p_1 & -\dfrac{p_{11}p_{12}}{2p_1p_2 - \frac{1}{2}p_{12}} \end{pmatrix}. \tag{5.18}$$

Then, the model of environmental/genetic effects can be expressed in the form $\mathbf{G} = \mathbf{SE}$ as:

$$\begin{pmatrix} G_{111} \\ G_{121} \\ G_{221} \\ G_{112} \\ G_{122} \\ G_{222} \end{pmatrix} = \begin{pmatrix} 1 & -2p_2 & -\dfrac{p_{12}p_{22}}{2p_1p_2 - \frac{1}{2}p_{12}} & -q_2 & 2p_2q_2 & \dfrac{p_{12}p_{22}q_2}{2p_1p_2 - \frac{1}{2}p_{12}} \\ 1 & p_1 - p_2 & \dfrac{p_{11}p_{22}}{p_1p_2 - \frac{1}{4}p_{12}} & -q_2 & (p_2 - p_1)q_2 & -\dfrac{p_{11}p_{22}q_2}{p_1p_2 - \frac{1}{4}p_{12}} \\ 1 & 2p_1 & -\dfrac{p_{11}p_{12}}{2p_1p_2 - \frac{1}{2}p_{12}} & -q_2 & -2p_1q_2 & \dfrac{p_{11}p_{12}q_2}{2p_1p_2 - \frac{1}{2}p_{12}} \\ 1 & -2p_2 & -\dfrac{p_{12}p_{22}}{2p_1p_2 - \frac{1}{2}p_{12}} & q_1 & -2p_2q_1 & -\dfrac{p_{12}p_{22}q_1}{2p_1p_2 - \frac{1}{2}p_{12}} \\ 1 & p_1 - p_2 & \dfrac{p_{11}p_{22}}{p_1p_2 - \frac{1}{4}p_{12}} & q_1 & (p_1 - p_2)q_1 & \dfrac{p_{11}p_{22}q_1}{p_1p_2 - \frac{1}{4}p_{12}} \\ 1 & 2p_1 & -\dfrac{p_{11}p_{12}}{2p_1p_2 - \frac{1}{2}p_{12}} & q_1 & 2p_1q_1 & -\dfrac{p_{11}p_{12}q_1}{2p_1p_2 - \frac{1}{2}p_{12}} \end{pmatrix} \begin{pmatrix} \mu \\ \alpha \\ \delta \\ \varepsilon \\ \alpha\varepsilon \\ \delta\varepsilon \end{pmatrix}. \tag{5.19}$$

The action of gene-environment interaction may thus be accounted for by two parameters—additive-environment and dominance-environment interactions, $\alpha\varepsilon$ and $\delta\varepsilon$, respectively. Ma et al. (2012) got to Expression 5.19, which can be easily generalized (by just keeping on adding the corresponding genetic- and/or environmental-effects design matrices with the Kronecker product) to more complex genetic architectures and several environmental factors—with increasingly possible gene-environment and also environment-environment interactions.

Additionally, the way we have developed the environmental part of the model above is straightforwardly extended to an arbitrary number of environmental alternatives at each environmental factor (and, thus, to a multiallelic haploid locus as well). By following the procedure above, it can be easily added, for instance, one additional environmental alternative, E_3, in Expressions 5.15–5.17, leading to the following model at the population level with three environmental alternatives:

$$\begin{pmatrix} G_1 \\ G_2 \\ G_3 \end{pmatrix} = \begin{pmatrix} 1 & -q_2 & -q_3 \\ 1 & 1-q_2 & -q_3 \\ 1 & -q_2 & 1-q_3 \end{pmatrix} \begin{pmatrix} \mu \\ \varepsilon^{12} \\ \varepsilon^{13} \end{pmatrix}, \tag{5.20}$$

where the variables of the environmental effects now indicate, in their superscripts, which pair of environmental alternatives stand for.

Overall, the results presented in this section enable us to model an arbitrary number of environmental factors with arbitrary environmental alternatives and arbitrary gene-environment interactions. In the same way as the use of the Kronecker product for developing multiple-locus models restricts the results to linkage equilibrium in the previous sections, the gene-environment interaction models developed in the current section are restricted to random associations between genotypes and environments. This may be regarded as a severe constraint, as there will exist cases with genotypes having differential preferences about the environments and/or selection acting in different ways on the genotypes under different environments. This constraint will be resolved in the next chapter.

5.6 Sex-Linked Loci

As in the previous sections of this chapter, we shall use the models for sex-linked loci at the individual level developed in the previous chapter as a mould to shape the corresponding models at the population level. We indeed start again with loci at the chromosome determining the homogametic sex, which we shall refer to as X-linked inheritance, in relation with the XY sex-determination system—although the developments will fit the Z-linked inheritance of the ZW sex-determination system just as well. Next, we shall address the case of Y-linked loci—also fitting the W-linked case.

5.6.1 X-Linked Loci

5.6.1.1 Separate Female and Male Genetic Effects

Let us consider a two-allele X-linked locus (with alleles X_1, X_2). One way of setting up Kempthorne's (1957) regression (see Expression 5.6) for an X-linked locus is $\mathbf{G} = \mathbf{N}_\mu\boldsymbol{\mu} + \mathbf{N}_\alpha\boldsymbol{\alpha} + \boldsymbol{\eta}$, expanding to:

$$\begin{pmatrix} G_{11} \\ G_{12} \\ G_{22} \\ G_1 \\ G_2 \end{pmatrix} = \begin{pmatrix} 1 & 0 \\ 1 & 0 \\ 1 & 0 \\ 0 & 1 \\ 0 & 1 \end{pmatrix} \begin{pmatrix} \mu_f \\ \mu_m \end{pmatrix} + \begin{pmatrix} 2 & 0 & 0 & 0 \\ 1 & 1 & 0 & 0 \\ 0 & 2 & 0 & 0 \\ 0 & 0 & 1 & 0 \\ 0 & 0 & 0 & 1 \end{pmatrix} \begin{pmatrix} \alpha_{f1} \\ \alpha_{f2} \\ \alpha_{m1} \\ \alpha_{m2} \end{pmatrix} + \begin{pmatrix} \eta_{11} \\ \eta_{12} \\ \eta_{22} \\ \eta_1 \\ \eta_2 \end{pmatrix}. \quad (5.21)$$

This proposal is the alternative at the population level to the setting at the individual level in Expressions 4.23 and 4.24. Indeed, it considers the effects of the two loci mainly as independent—to the point of having different referent points, μ_f and μ_m. Besides, matrix $\mathbf{N}_\alpha$ is block diagonal, with a first block assigning the female genetic effects to the female genotypes and a second (trivial) block assigning the male genetic effects to the male genotypes.

However, the population level establishes a link between the parameters of the two sexes, since the population frequencies are here held within the same basket, i.e., $p_{f11} + p_{f12} + p_{f22} + p_{m1} + p_{m2} = 1$. Indeed, clearing away the reference point is for this case—with actually two reference points—slightly trickier than in the previous cases (for which we just used Expression 5.7). For the present case, we have to solve $\mathbf{G} = \mathbf{N}_\mu\boldsymbol{\mu} + \boldsymbol{\eta}_\mu$. Following the indications given above (Expressions 5.7–5.9 and related text), it is easy to obtain $\boldsymbol{\mu} = \tilde{\mathbf{H}}_\mu$, with $\tilde{\mathbf{H}}_\mu = (\mathbf{N}_\mu^T\mathbf{P}\mathbf{N}_\mu)^{-1}\mathbf{N}_\mu^T\mathbf{P}\mathbf{G}$. This leads to $\mu_f = p_{f11}\,G_{11} + p_{f12}\,G_{12} + p_{f22}\,G_{22}$ and $\mu_m = p_{m1}\,G_1 + p_{m2}\,G_2$. These parameters would be the female and male population means, respectively, as long as $p_{f11} + p_{f12} + p_{f22} = 1$ and $p_{♂1} + p_{♂2} = 1$. That is however not the case, as mentioned just above. Nevertheless, the proper female and male population means can be obtained from them just as $\mu_f/(p_{f11} + p_{f12} + p_{f22})$ and $\mu_m/(p_{m1} + p_{m2})$, respectively. Thus, what the regression in Expression 5.21 does is to split the population mean into a female and a male components, $\mu = \mu_f + \mu_m$, which are used as partial reference points.

Then, following again the indications in Expressions 5.7–5.9 and related text, it is possible to obtain the error term of the first step of regression in Expression 5.21—i.e., of regression $\mathbf{G} = \mathbf{N}_\mu\boldsymbol{\mu} + \boldsymbol{\eta}_\mu$—is $\boldsymbol{\eta}_\mu = \mathbf{I} - \mathbf{N}_\mu\tilde{\mathbf{H}}_\mu\mathbf{G}$. With this error term as dependent variable, we can set the next step of regression in Expression (5.21)—i.e., regression $\boldsymbol{\eta}_\mu = \mathbf{N}_\alpha\boldsymbol{\alpha} + \boldsymbol{\eta}$—and solve it following the same steps as in the previous cases. In this way, we get to the solution of Kempthorne's (1957) regression for the case in question (Expression 5.21) and, therefore, to build the genetic-effects design matrix as in previous sections. In particular, in the present case, the expression of the kind $\mathbf{E} = \mathbf{S}^{-1}\mathbf{G}$ expands to:

$$\begin{pmatrix} \mu_f \\ \alpha_f \\ \delta_f \\ \mu_m \\ \alpha_m \end{pmatrix} = \begin{pmatrix} p_{f11} & p_{f12} & p_{f22} & 0 & 0 \\ -\frac{p_{f11}p_{f2}}{2p_{f1}p_{f2} - \frac{1}{2}p_{f12}(p_{f1}+p_{f2})} & \frac{p_{f12}(p_{f1}-p_{f2})}{4p_{f1}p_{f2} - p_{f12}(p_{f1}+p_{f2})} & \frac{p_{f1}p_{f22}}{2p_{f1}p_{f2} - \frac{1}{2}p_{f12}(p_{f1}+p_{f2})} & 0 & 0 \\ -\frac{1}{2} & 1 & -\frac{1}{2} & 0 & 0 \\ 0 & 0 & 0 & p_{m1} & p_{m2} \\ 0 & 0 & 0 & -1 & 1 \end{pmatrix} \begin{pmatrix} G_{11} \\ G_{12} \\ G_{22} \\ G_1 \\ G_2 \end{pmatrix} \tag{5.22}$$

The matrix in this expression is a block-diagonal matrix with two blocks—an upper-left block for the female part and a bottom-right one for the male part— which actually comes from $\mathbf{N}_\mu$ and $\mathbf{N}_\alpha$ in Expression 5.21 also being block-diagonal matrices. This, in turn, is a consequence of the genetic-effects design matrix of the system at the individual level (Expressions 5.23, 5.24) also being a block diagonal matrix.

By comparing the upper-left female block in Expression 5.22 to the initial one-locus two-allele autosomal gene (Expression 5.11), it is easy to see that the only difference between them is a factor $(p_{f1} + p_{f2})$. This factor actually equals one in the autosomal case. Thus, the female block of the X-linked case and the matrix of the autosomal case are different only in what regards the female frequencies of the second case being only a subset of all the frequencies of the genetic system (which includes the male frequencies as well). The bottom-down male block, in its turn, is the same as the matrix of an environmental factor alone—i.e., of a haploid locus, which is in fact the way the locus behaves in males.

The matrix in Expression 5.22 can be inverted by separately inverting its two diagonal blocks, leading to the equivalent expression $\mathbf{G} = \mathbf{SE}$. Thus, the resulting genetic-effects design matrix, $\mathbf{S}$, is also a block-diagonal matrix with an upper-left 3×3 female block and a bottom-right 2×2 male block. However, the simplification of the formulae becomes for this case more limited than in the cases equivalent to each block—a diploid biallelic autosomal locus and a haploid biallelic locus, respectively. For this reason, the expanded form of the $\mathbf{G} = \mathbf{SE}$ expression for the general case (i.e., for any set of genotype frequencies) becomes already cumbersome enough not to be worthwhile to be made explicit here, given that the way such expression may be obtained numerically for a particular case is explained right above.

The extension of this model to multiple alleles can be obtained by just enlarging Expression 5.21 to accommodate them. For three alleles, for instance, the two diagonal blocks of matrix $\mathbf{N}_\alpha$ would be matrix $\mathbf{N}$ in Expression 5.14 and an identity matrix of dimension 3. Whether larger, the solution of the regression framework and the $\mathbf{E} = \mathbf{S}^{-1}\mathbf{G}$ and $\mathbf{G} = \mathbf{SE}$ expressions can be obtained following the indications for the biallelic case and considering also the indications given for the case of a diploid autosomal multiallelic locus (Expression 5.14 and related text) and for an haploid multiallelic locus (Expression 5.20 and related text).

5.6.1.2 The Difference Between Females and Males Made Explicit

As mentioned in the previous chapter on models at the individual level, there is an alternative way of modeling an X-linked locus, which explicitly accounts for the putative difference between homogytous (female) and hemizygous (male) genotypes. The population-level version of that setting shall thus enable direct testing on whether those genotypes lead to the same phenotypes, as it happens in some cases, including the first case reported of an X-linked locus (Morgan 1910).

Such model can be obtained from regression framework $\mathbf{G} = \mathbf{1}\,\mu + \mathbf{N}_\alpha\boldsymbol{\alpha} + \mathbf{N}_\delta\boldsymbol{\delta} + \boldsymbol{\eta}$, expanding to:

$$\begin{pmatrix} G_{11} \\ G_{12} \\ G_{22} \\ G_1 \\ G_2 \end{pmatrix} = \begin{pmatrix} 1 \\ 1 \\ 1 \\ 1 \\ 1 \end{pmatrix} \mu + \begin{pmatrix} 2 & 0 \\ 1 & 1 \\ 0 & 2 \\ 2 & 0 \\ 0 & 2 \end{pmatrix} \begin{pmatrix} \alpha_1 \\ \alpha_2 \end{pmatrix} + \begin{pmatrix} 1 & 0 & 0 \\ 0 & 1 & 0 \\ 0 & 0 & 1 \\ 1 & 0 & 0 \\ 0 & 0 & 1 \end{pmatrix} \begin{pmatrix} \delta_{11} \\ \delta_{12} \\ \delta_{22} \end{pmatrix} + \begin{pmatrix} \eta_{11} \\ \eta_{12} \\ \eta_{22} \\ \eta_1 \\ \eta_2 \end{pmatrix}. \tag{5.23}$$

This regression framework expresses the dependent variable, $\mathbf{G}$, in terms of three explanatory variables, μ, $\boldsymbol{\alpha}$, and $\boldsymbol{\delta}$, plus an error term, $\boldsymbol{\eta}$. Design matrix $\mathbf{N}_\alpha$ assigns common additive effects to males and females—it assigns two additive effects to homozygous (female) genotypes (with two alleles), as well as to hemizygous (male) genotypes, although the latest ones have one only allele. Thus, the additive effect in a hemizygous (male) genotype has double the effect than in a homozygous (female) genotype. This is indeed in accordance with the aforementioned by-default assumption that homozygous and their corresponding hemizygous genotypes, with different allele content, may lead to the same phenotype.

As in the previous cases, regression framework in Expression 5.23 is to be solved sequentially—the residual term of one regression provides the dependent variable of the following one—with, this time, up to three sequential steps. The regression steps for this case are, in particular, $\mathbf{G} = \mathbf{1}\,\mu + \boldsymbol{\eta}_\mu$, $\boldsymbol{\eta}_\mu = \mathbf{N}_\alpha\boldsymbol{\alpha} + \boldsymbol{\eta}_\alpha$, and $\boldsymbol{\eta}_\alpha = \mathbf{N}_\delta\boldsymbol{\delta} + \boldsymbol{\eta}$. All regressions can be solved with the method described in Expressions 5.7–5.9 and related text. As in all previous cases but the latest one, the first regression is trivial, just leading to the mean, $\mu = \mathbf{1}^{\mathrm{T}}\mathbf{PG}$, and to the mean-corrected genotypic values, $\boldsymbol{\eta}_\mu = \overline{\mathbf{G}} = \mathbf{G} - \mathbf{1}\,\mu$.

With this, we get the population mean phenotype, μ, the additive effects, $\boldsymbol{\alpha}$, and the dominance effects, $\boldsymbol{\delta}$, which enable us to build the inverse of the genetic-effects design matrix, $\mathbf{S}^{-1}$, providing the genetic effects at the population level, $\mathbf{E} = (\mu, \alpha, \delta, \xi_1, \xi_2)^{\mathrm{T}}$, in terms of the genotypic values, $\mathbf{G} = (G_{11}, G_{12}, G_{22}, G_1, G_2)^{\mathrm{T}}$, i.e., $\mathbf{E} = \mathbf{S}^{-}$

^{1}**G**. In particular, the first three rows of $\mathbf{S}^{-1}$ is given by the scalars of μ, $\alpha = \alpha_2 - \alpha_1$, and $\delta = \delta_{12} + (\delta_{11} + \delta_{22})/2$, in terms of **G**. The remaining two rows are the same as at the individual level (Expression 4.26), i.e., $(-1, 0, 0, 1, 0)$ and $(0, 0, -1, 0, 1)$. The equivalent expression $\mathbf{G} = \mathbf{SE}$ can then be easily obtained by just obtaining the genetic-effects design matrix, **S**, as the inverse of $\mathbf{S}^{-1}$.

It is easy to extend Expression 5.23 to several alleles. Matrix $\mathbf{N}_\alpha$ would keep on having an upper block corresponding to the female genotypic values that can be extended to multiple alleles just as in the case of an autosomal locus—for three alleles, for instance, see **N** in Expression 5.14—and a lower diagonal block corresponding to the male genotypic values, in which all scalars at the diagonal are "2". Matrix $\mathbf{N}_\delta$ has also an upper and a lower block for females and males, respectively. The upper block is an identity matrix and the lower block is the result of removing, from that matrix, the rows of the heterozygotes. Then, the multiple-alleles $\mathbf{S}^{-1}$ matrix can be built from the result of the regression following the same logic as in the two-locus case, with just an increased number of genetic effects. For three alleles, for instance, the model would include two additive effects, $\alpha^{12} = \alpha_2 - \alpha_1$ and $\alpha^{13} = \alpha_3 - \alpha_1$. The additive genetic effect $\alpha^{23} = \alpha_3 - \alpha_2$ also exists but, as in the case of an autosomal three-allele locus, it needs not be included in the model as it can be retrieved whenever necessary from the genetic effects included in the model as $\alpha^{23} = \alpha^{13} - \alpha^{12}$.

Overall, this alternative way of modeling an X-linked locus provides a population mean and additive effects that are averaged over the whole population—thus including both females and males. Although there are no dominance effects by themselves in (hemizygous) males, the male genotypic values are taken into account for wcomputing the dominance effects—as it can be seen in the two last rows of $\mathbf{N}_\delta$ (Expression 5.23). Indeed, in what regards the population mean and the additive and dominance effects, the model considers the male hemyzygous and the female homozygous genotypes in the very same way—cf. rows 1 and 4 and rows 3 and 5 of $\mathbf{N}_\alpha$ (Expression 5.23). Then, the model provides explicit parameters for accounting for the putative difference between the homozygous (female) and the hemizygous (male) genotypic values, ξ_1 and ξ_2. Since these estimates shall naturally come with errors when obtained from a dataset, they can be used directly to test whether the homozygous and the hemizygous genotypes are significantly different or not.

5.6.1.3 Y-Linked Loci

As for the individual level (in the previous chapter), models for Y-linked loci at the population level are valid for Z-linked loci in species with the ZW sex-determination system. Also analogous to the individual level, the expressions developed for integrating an environmental factor with two environmental alternatives into a gene-environment interaction case (Expressions 5.16, 5.17) are just the ones to be used to model a two-allele Y-linked locus at the population level. The extension to multiple loci is also done in the very same way as the extension of the environmental factor to more environmental alternatives (Expression 5.20).

Once again as commented for Y-linked loci at the individual level in the previous chapter, we might be interested in Y-linked loci affecting traits that are also affected by autosomal loci as well and also displayed in females (see Expressions 4.27 and 4.28 and related text in Chap. 4). Such model is very simple, actually analogous to the one developed above for an environmental variable (Expressions 5.15–5.17). For the present case, we consider regression $\mathbf{G} = \mathbf{1}\,\mu + \boldsymbol{\alpha}$, expanding to:

$$\begin{pmatrix} G_1 \\ G_2 \\ G_f \end{pmatrix} = \begin{pmatrix} 1 \\ 1 \\ 1 \end{pmatrix} \mu + \begin{pmatrix} \alpha_1 \\ \alpha_2 \\ \beta \end{pmatrix}. \tag{5.24}$$

The solution to this regression just provides the population average phenotype. Then, we define $\alpha = \alpha_2 - \alpha_1$, just as in the one-locus two-allele autosomal case (Expression 5.6 and related text), and $\varphi = \beta - (\alpha_1 + \alpha_2)/2$, for consistency with the aforementioned model for and Y-linked locus at the individual level. With this, we can obtain the inverse of the genetic-effects design matrix and the expression $\mathbf{E} = \mathbf{S}^{-1}\mathbf{G}$ as:

$$\begin{pmatrix} \mu \\ \alpha \\ \varphi \end{pmatrix} = \begin{pmatrix} p_1 & p_2 & p_f \\ -1 & 1 & 0 \\ -½ & -½ & 1 \end{pmatrix} \begin{pmatrix} G_1 \\ G_2 \\ G_f \end{pmatrix}. \tag{5.25}$$

By comparing this expression with its equivalent at the individual level (Expression 4.28), it can be seen that the only different row is the first one, providing the mean, which can be deduced without the need of a regression—particularly, of regression in Expression 5.24.

In any case, by inverting matrix $\mathbf{S}^{-1}$ in Expression 5.25, we can obtain the equivalent expression $\mathbf{G} = \mathbf{S}^{-1}\mathbf{E}$, as:

$$\begin{pmatrix} G_1 \\ G_2 \\ G_f \end{pmatrix} = \begin{pmatrix} 1 & -\left(p_2 + \frac{p_f}{2}\right) & -p_f \\ 1 & p_1 + \frac{p_f}{2} & -p_f \\ 1 & p_1 - p_2 & p_1 + p_2 \end{pmatrix} \begin{pmatrix} \mu \\ \alpha \\ \varphi \end{pmatrix}. \tag{5.26}$$

The extension of the model to several alleles is straightforward particularly from Expression 5.25. Indeed, the k-allele $\mathbf{S}^{-1}$ matrix has a first row with the frequencies and the remaining rows as the equivalent matrix at the individual level (see Expression 4.28 and related text in Chap. 4).

It is worthwhile to keep in mind that the additive variance for this system can be obtained using the standard procedure (see Expression 2.11 and precedent text in Chap. 2) as $V_A = p_1\alpha_1^2 + p_2\alpha_2^2$ for the two-locus case and $V_A = \sum_{i=1}^{k} p_i\alpha_i^2$, for the general case. This is to say, the females do not contribute to the additive variance of a Y-linked locus because they do not bear any allele of the locus in question. Precisely to avoid confusion to this regard, we have used β, instead of e.g., α_f, to name the last scalar in vector $\boldsymbol{\alpha}$ (Expression 5.24). Also keep in mind that female parameters are included in this one-locus model to enable it to be implemented into a more complex genetic architecture, including also autosomal loci, as already mentioned in the previous chapter.

5.7 Gene-Sex Interaction

A locus whose genotypes underlay different phenotypes in females and males can be modeled with the same two alternative perspectives followed above to model an X-linked locus—i.e., with a set of genetic effects for each sex or with a common set of genetic effects for the two sexes plus a set of genetic effects accounting for the sex differences. For the first case, Kempthorne's (1957) regression locus is $\mathbf{G} = \mathbf{N}_{\mu}\boldsymbol{\mu} + \mathbf{N}_{\alpha}\boldsymbol{\alpha} + \boldsymbol{\eta}$, expanding to:

$$\begin{pmatrix} G_{f11} \\ G_{f12} \\ G_{f22} \\ G_{m11} \\ G_{m12} \\ G_{m22} \end{pmatrix} = \begin{pmatrix} 1 & 0 \\ 1 & 0 \\ 1 & 0 \\ 0 & 1 \\ 0 & 1 \\ 0 & 1 \end{pmatrix} \begin{pmatrix} \mu_f \\ \mu_m \end{pmatrix} + \begin{pmatrix} 2 & 0 & 0 & 0 \\ 1 & 1 & 0 & 0 \\ 0 & 2 & 0 & 0 \\ 0 & 0 & 2 & 0 \\ 0 & 0 & 1 & 1 \\ 0 & 0 & 0 & 2 \end{pmatrix} \begin{pmatrix} \alpha_{f1} \\ \alpha_{f2} \\ \alpha_{m1} \\ \alpha_{m2} \end{pmatrix} + \begin{pmatrix} \eta_{f11} \\ \eta_{f12} \\ \eta_{f22} \\ \eta_{m11} \\ \eta_{m12} \\ \eta_{m22} \end{pmatrix}. \quad (5.27)$$

Indeed, although slightly larger, this expression can be considered simpler than Expression 5.21, in which females are diploid and males are haploid—in this case, both sexes are equally diploid.

Following the same steps as in the case of an X-linked locus, the solution to Kempthorne's (1957) regression for this case (Expression 5.27) leads to an inverse of the genetic-effects design matrix with two diagonal blocks. In this case, the bloc of males is the same as the one of females, with just male frequencies, p_{mij} and p_{mi}, instead of female frequencies, p_{fij} and p_{fi} (cf. Expression 5.22). With this, we obtain the expression $\mathbf{E} = \mathbf{S}^{-1}\mathbf{G}$ and, from it, its equivalent expression $\mathbf{G} = \mathbf{SE}$. The extension to several alleles can be obtained by just enlarging Expression 5.27 to accommodate multiple alleles both in females and in males, as in the case of an X-linked locus above (Expressions 5.22 and 5.23 and related text).

Concerning now the option of common genetic effects for the two sexes and additional genetic effects explicitly accounting for the differences, as explained for the models at the individual level (Expressions 4.30 and 4.21 and related text in Chap. 4), gene-sex interaction works mathematically in the same way as gene-environment interaction. Thus, the developments made above for gene-environment interaction (Expressions 5.16–5.20 and related text), apply to this case by just replacing the environmental effect, ε, by the sex effect, σ. Incidentally, Expression 5.27 is an alternative way to model gene-environment interactions, by just replacing the subscripts, f and m, to design environments instead.

Then, as explained for the gene-environment interaction case above, the developments in this chapter apply to the case when random associations occur. In the present case of gene-sex interaction modeled with explicit effects accounting for the differences between the sexes, this means that the model assumes that the genotypes are equally distributed across the two sexes. As in any of the previous cases, the present gene-sex interaction one-locus models can be implemented into more complex genetic architectures using the Kronecker product of genetic-effects design matrices (or their inverses), in which case the assumption of random associations implies that the frequencies are assumed to be under linkage equilibrium.

5.8 Imprinting

To end up with, we consider the case of an imprinted locus with, initially, two alleles, A_1 and A_2. In contrast to the case of a non-imprinted locus, we now need to consider four different genotypes (instead of just three), A_1A_1, A_1A_2, A_2A_1, and A_2A_2. As explained in the previous chapter for the individual level, modeling this situation also allows two different perspectives. On the one hand, we can consider a model with two different dominance effects. On the other hand, we can consider a common (averaged) dominance effect plus an explicit imprinting effect.

5.8.1 Two Dominance Effects

For the option of two dominance effects, Kempthorne's (1957) regression can be obtained as a straightforward extension of Expression 5.6, with the same general setting, $\mathbf{G} = \mathbf{1}\,\mu + \mathbf{N}\boldsymbol{\alpha} + \boldsymbol{\delta}$, but this time expanding to:

$$\begin{pmatrix} G_{11} \\ G_{12} \\ G_{21} \\ G_{22} \end{pmatrix} = \begin{pmatrix} 1 \\ 1 \\ 1 \\ 1 \end{pmatrix} \mu + \begin{pmatrix} 2 & 0 \\ 1 & 1 \\ 1 & 1 \\ 0 & 2 \end{pmatrix} \begin{pmatrix} \alpha_1 \\ \alpha_2 \end{pmatrix} + \begin{pmatrix} \delta_{11} \\ \delta_{12} \\ \delta_{21} \\ \delta_{22} \end{pmatrix}. \tag{5.28}$$

With the solution of this regression (obtained, as in the previous cases, using Expressions (5.7–5.9) above), we define (as in the non-imprinted case above) $\alpha = \alpha_2 - \alpha_1$ and thus obtain the expression $\mathbf{E} = \mathbf{S}^{-1}\mathbf{G}$ as:

$$\begin{pmatrix} \mu \\ \alpha \\ \delta^{12} \\ \delta^{21} \end{pmatrix} = \begin{pmatrix} p_{11} & p_{12} & p_{21} & p_{22} \\ -\dfrac{p_{11}p_2}{p_{11}p_2+p_1p_{22}} & \dfrac{p_{12}(p_1-p_2)}{2(p_{11}p_2+p_1p_{22})} & \dfrac{p_{21}(p_1-p_2)}{2(p_{11}p_2+p_1p_{22})} & \dfrac{p_1p_{22}}{p_{11}p_2+p_1p_{22}} \\ -\frac{1}{2} & 1 & 0 & -\frac{1}{2} \\ -\frac{1}{2} & 0 & 1 & -\frac{1}{2} \end{pmatrix} \begin{pmatrix} G_{11} \\ G_{12} \\ G_{21} \\ G_{22} \end{pmatrix} \tag{5.29}$$

As usual, the rows of the upper-level genetic effects at the bottom of the inverse of the genetic-effects design matrix, $\mathbf{S}^{-1}$ (in this case, the two dominance effects), coincide with the corresponding expression at the individual level (Expression 4.33). Then, by inverting $\mathbf{S}^{-1}$ in Expression 5.29, we obtain $\mathbf{G} = \mathbf{SE}$ as:

$$\begin{pmatrix} G_{11} \\ G_{12} \\ G_{21} \\ G_{22} \end{pmatrix} = \begin{pmatrix} 1 & -2p_2 & -\dfrac{p_{12}p_{22}}{p_{11}p_2 + p_1p_{22}} & -\dfrac{p_{21}p_{22}}{p_{11}p_2 + p_1p_{22}} \\ 1 & p_1 - p_2 & \dfrac{4p_{11}p_{22} + p_{21}(p_{11} + p_{22})}{2(p_{11}p_2 + p_1p_{22})} & -\dfrac{p_{21}(p_{11} + p_{22})}{2(p_{11}p_2 + p_1p_{22})} \\ 1 & p_1 - p_2 & -\dfrac{p_{12}(p_{11} + p_{22})}{2(p_{11}p_2 + p_1p_{22})} & \dfrac{4p_{11}p_{22} + p_{12}(p_{11} + p_{22})}{2(p_{11}p_2 + p_1p_{22})} \\ 1 & 2p_1 & -\dfrac{p_{11}p_{12}}{p_{11}p_2 + p_1p_{22}} & -\dfrac{p_{11}p_{21}}{p_{11}p_2 + p_1p_{22}} \end{pmatrix} \begin{pmatrix} \mu \\ \alpha \\ \delta^{12} \\ \delta^{21} \end{pmatrix}. \tag{5.30}$$

The denominators of Expressions 5.29 and 5.30 can be expressed in a way very similar to the non-imprinted case (Expressions 5.11, 5.12) since with some algebra it is possible to prove that, for the present case, $p_{11}p_2 + p_{22}p_1 = 2p_1p_2 - ½(p_{12} + p_{21})$. Incidentally, due to a regrettable mistake, these denominators were disclosed wrongly in the original publication (Expression 7 in Álvarez-Castro 2014). In that publication, the model was developed directly by solving a regression of the type described by Fisher (1918) (see Fig. 5.3), which is further explained in the discussion below. Such approach and the one followed above lead to the same result (Expressions 5.29, 5.30). This is to point out, it was just by an unfortunate mistake in polishing its final form—and not due to the approach followed—that Expression 7 in Álvarez-Castro (2014) was not correct.

5.8.2 Explicit Imprinting Effect

As mentioned above, we can alternatively opt for a model with an explicit imprinting parameter, ι. Thus, such model entails both additive, dominance, and imprinting components. The regression framework for that model is $\mathbf{G} = \mathbf{1}\,\mu + \mathbf{N}_{\alpha}\boldsymbol{\alpha} + \mathbf{N}_{\delta}\boldsymbol{\delta} + \boldsymbol{\eta}$, expanding to:

$$\begin{pmatrix} G_{11} \\ G_{12} \\ G_{21} \\ G_{22} \end{pmatrix} = \begin{pmatrix} 1 \\ 1 \\ 1 \\ 1 \end{pmatrix} \mu + \begin{pmatrix} 2 & 0 \\ 1 & 1 \\ 1 & 1 \\ 0 & 2 \end{pmatrix} \begin{pmatrix} \alpha_1 \\ \alpha_2 \end{pmatrix} + \begin{pmatrix} 1 & 0 & 0 \\ 0 & 1 & 0 \\ 0 & 1 & 0 \\ 0 & 0 & 1 \end{pmatrix} \begin{pmatrix} \delta_{11} \\ \delta_{12} \\ \delta_{22} \end{pmatrix} + \begin{pmatrix} \eta_{11} \\ \eta_{12} \\ \eta_{21} \\ \eta_{22} \end{pmatrix}. \tag{5.31}$$

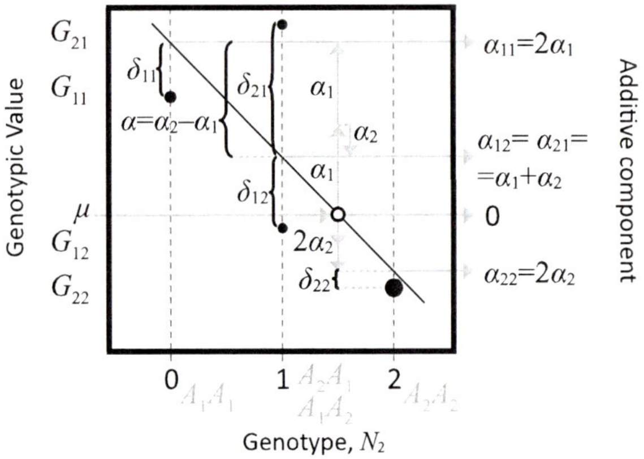

Fig. 5.3 Graphical interpretation of Fisher's (1918) additive and dominance population parameters, for the case of an imprinted locus. As in the non-imprinted case (see Fig. 5.1), the additive parameters are the average effects of allele substitutions, α_i, $i = 1, 2$. With imprinting, we can consider four breeding values $\alpha_{ij} = \alpha_i + \alpha_j$, $ij = 11, 12, 21, 22$—i.e., one additional heterozygous breeding value—but the two heterozygous breeding values are actually equal. There is also one additional dominance parameter as compared with the non imprinted case—the four dominance parameters are δ_{ij}, $ij = 11, 12, 21, 22$—but now the two heterozygous dominance parameters do differ. Indeed, if the two dominance parameters are not different, the locus is non-imprinted. As in the non-imprinted case, the genotypes are represented in the horizontal axis of the graph by their content of allele A_2, N_2, and the black solid line is the regression (linear approximation) to the genotypic values

This regression has to be solved through three sequential steps, analogous to the second proposal for the case of a sex-linked locus above (see Expression 5.23 and related text).

The first two regression steps of Expression 5.31, providing the mean phenotype and the additive component, are the same as in the previous model (Expression 5.28). Therefore, the first two rows of the inverse of the genetic-effects design matrix, $\mathbf{S}^{-1}$, are the same as those of the corresponding matrix of the previous model (Expression 5.29). The third step of the regression provides $\boldsymbol{\delta}$, from which the dominance genetic effect (and, hence, the third row of $\mathbf{S}^{-1}$) can be obtained as $\delta = \delta_{12} - (\delta_{11} + \delta_{22})/2$. Since the fourth row of $\mathbf{S}^{-1}$ is also the last one, it coincides with that of the corresponding matrix at the individual level (Expression 4.35). On the whole, the model can be expressed as $\mathbf{E} = \mathbf{S}^{-1}\mathbf{G}$, expanding to:

$$
\begin{pmatrix} \mu \\ \alpha \\ \delta \\ \iota \end{pmatrix}
= \begin{pmatrix}
p_{11} & p_{12} & p_{21} & p_{22} \\
-\dfrac{p_{11}p_2}{p_{11}p_2+p_1p_{22}} & \dfrac{p_{12}(p_1-p_2)}{2(p_{11}p_2+p_1p_{22})} & \dfrac{p_{21}(p_1-p_2)}{2(p_{11}p_2+p_1p_{22})} & \dfrac{p_1p_{22}}{p_{11}p_2+p_1p_{22}} \\
-½ & \dfrac{p_{12}}{p_{12}+p_{21}} & \dfrac{p_{21}}{p_{12}+p_{21}} & -½ \\
0 & -½ & ½ & 0
\end{pmatrix}
\begin{pmatrix} G_{11} \\ G_{12} \\ G_{21} \\ G_{22} \end{pmatrix}. \tag{5.32}
$$

As in the previous cases, the model in the form $\mathbf{G} = \mathbf{SE}$ comes from inverting the previously derived matrix $\mathbf{S}^{-1}$. This way it is possible to obtain (Álvarez-Castro 2014):

$$
\begin{pmatrix} G_{11} \\ G_{12} \\ G_{21} \\ G_{22} \end{pmatrix}
= \begin{pmatrix}
1 & -2p_2 & -\dfrac{p_{22}(p_{12}+p_{21})}{p_{11}p_2+p_1p_{22}} & 0 \\
1 & p_1-p_2 & \dfrac{2p_{11}p_{22}}{p_{11}p_2+p_1p_{22}} & -\dfrac{2p_{21}}{p_{12}+p_{21}} \\
1 & p_1-p_2 & \dfrac{2p_{11}p_{22}}{p_{11}p_2+p_1p_{22}} & \dfrac{2p_{12}}{p_{12}+p_{21}} \\
1 & 2p_1 & -\dfrac{p_{11}(p_{12}+p_{21})}{p_{11}p_2+p_1p_{22}} & 0
\end{pmatrix}
\begin{pmatrix} \mu \\ \alpha \\ \delta \\ \iota \end{pmatrix}. \tag{5.33}
$$

Using this option for modeling imprinting, the estimate of the imprinting genetic effect, ι (together with its standard error), directly provides a test for imprinting. Using the previous option, with two dominance effects, an equivalent test can be performed by testing whether the two dominance effects are significantly different. Nevertheless, the present option with an explicit imprinting effect also enables a decomposition of the genetic variance with an explicit imprinting component.

Indeed, the explicit imprinting effect in Expression 5.33 enables an orthogonal decomposition of the genotypic values and, thus, also of the genetic variance with an explicit imprinting component. Thus, following the same logic as in the section on orthogonal variance decomposition above, and assuming for simplicity the Hardy-Weinberg proportions, from Expression 5.33, it is possible to derive $V_A = 2p_1p_2(a + d(p_2 - p_1))^2$, $V_D = (2dp_1p_2)^2$, and $V_O = 2\iota^2 p_1p_2$ (Álvarez-Castro 2014). These expressions coincide with the well-known expressions of the non-imprinted case (Expressions 2.11 and 2.14, respectively), in what regards the additive and dominance components of the genotypic variance. Thus, an additional, orthogonal imprinting component is just added up on top of them.

Álvarez-Castro (2014) showed how to obtain a general (not dependent upon equilibrium frequencies) expression of the variance decomposition, using Álvarez-Castro and Yang's (2011) procedure for variance decompositions with NOIA, which

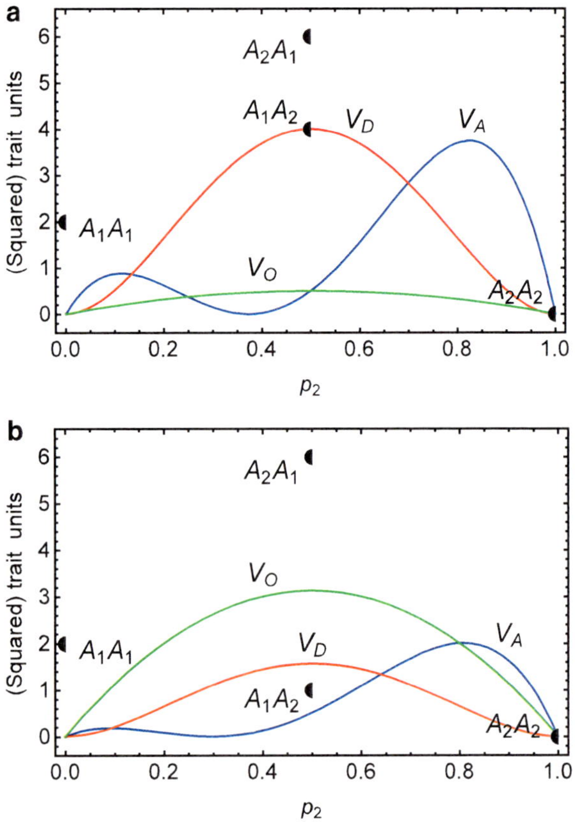

Fig. 5.4 Modified from Álvarez-Castro (2014). Decomposition of the genetic variance for one-locus two-allele cases with imprinting, for the whole range of possible frequencies (p_2 is the frequency of A_2), assuming Hardy-Weinberg. The additive variance (V_A, blue line), the dominance variance (V_D, red line), and the imprinting variance (V_O, green line) are measured in squared trait units. The genotypic values (black dots) are measured in trait units. (**a**) Additive and dominance genetic effects at the individual level (also in trait units) as in Fig. 2.6b (i.e., $a = -1$ and $d = 4$), plus an additional imprinting effect at the individual level, $i = -1$. (**b**) Case considering a larger imprinting effect, with $a = -1$, $d = 2.5$, and $i = -2.5$

shall be discussed in Chap. 7. Figure 5.4 shows two cases of variance decomposition with imprinting. Figure 5.4a takes up the case dealt with in Fig. 2.6b and adds imprinting to it—by just splitting the genotypic value of the heterozygote into two different imprinted genotypic values. As expected (from what has been discussed in the previous paragraph), V_A and V_D coincide in those two panels. In Fig. 5.4b, a case with more extreme imprinting is considered, showing that the imprinting variance may dominate the variance decomposition. A comprehensive discussion on the interpretation of the decomposition of the genetic variance is provided in Chap. 7.

Unfortunately, it has been wrongly claimed that Expression 5.33, originally published by Álvarez-Castro (2014), is not orthogonal (Deng et al. 2020). Thus, in order to dispel any possible doubts about the correctness of this expression, the matrix that results from applying the test of orthogonality (Expression 5.1) to Expression 5.33 is here provided as $\mathbf{S}^{\mathrm{T}}\mathbf{PS} = Diag(1, -p_{11}p_2 + p_{22}p_1, \frac{4p_{22}(p_2(2p_2-1)+p_{22}(1-3p_2+p_{22}))}{p_{11}p_2+p_{22}p_1}, \frac{4p_{12}p_{21}}{p_{12}+p_{21}})$. Since this is a diagonal matrix, the model in Expression 5.33, originally published by Álvarez-Castro (2014), is indeed orthogonal for any possible values the genotypic frequencies may take.

At any rate, both regression in Expression 5.28 of the two-dominance perspective and regression in Expression 5.31 of the explicit-imprinting-effect perspective can be extended to account for multiple alleles in a way analogous to the non-imprinted case above (Expressions 5.6–5.14 and related text). For instance, the regression for modeling an imprinted locus with three alleles under the two-dominance perspective is $\mathbf{G} = \mathbf{1}\,\mu + \mathbf{N}\boldsymbol{\alpha} + \boldsymbol{\delta}$, expanding to:

$$\begin{pmatrix} G_{11} \\ G_{12} \\ G_{21} \\ G_{22} \\ G_{13} \\ G_{31} \\ G_{23} \\ G_{32} \\ G_{33} \end{pmatrix} = \begin{pmatrix} 1 \\ 1 \\ 1 \\ 1 \\ 1 \\ 1 \\ 1 \\ 1 \\ 1 \end{pmatrix} \mu + \begin{pmatrix} 2 & 0 & 0 \\ 1 & 1 & 0 \\ 1 & 1 & 0 \\ 0 & 2 & 0 \\ 1 & 0 & 1 \\ 1 & 0 & 1 \\ 0 & 1 & 1 \\ 0 & 1 & 1 \\ 0 & 0 & 2 \end{pmatrix} \begin{pmatrix} \alpha_1 \\ \alpha_2 \\ \alpha_3 \end{pmatrix} + \begin{pmatrix} \delta_{11} \\ \delta_{12} \\ \delta_{21} \\ \delta_{22} \\ \delta_{13} \\ \delta_{31} \\ \delta_{23} \\ \delta_{32} \\ \delta_{33} \end{pmatrix}. \tag{5.34}$$

Also as in the previous cases, imprinted loci can be implemented in the context of a more complex genetic architecture, with multiple loci and/or gene-environment interactions, with the Kronecker product of the (inverses of the) genetic-effects design matrices of all the underlying factors of the trait, assuming linkage equilibrium.

5.9 Genetic Effects in Current Quantitative Genetics

In this chapter, we have worked out at the population level all the genetic architectures considered in the previous chapter at the individual level. The models here presented express the genotypic values in terms of the genetic effects, which are the variables needed to describe the quantitative genetics properties of a population. Models at the population level are aimed to be general not only in what regards genetic architectures—as the models at the individual level also are—but concerning complex population facts as well. In this regard, the developments at the population level here provided are correct—they reflect the biological meaning Fisher (1918) originally conceived for them to be useful quantitative genetic indexes—unless several factors (either loci alone or loci and environments) involving any kind of non-random association between/among them are considered. Such associations include departures from linkage equilibrium frequencies (implying only loci) and non-random associations between/among genotypes and environments (implying loci and environments). Those situations have been saved to be solved in the next chapter of this book.

5.9.1 New Developments

Some of the above developments are actually published for the first time in the present chapter—to the best of the author's knowledge. This is the case of the X-linked models and the gene-sex interaction models (Expressions 5.21–5.27). Also many of the indications to extend the different models to multiple alleles (e.g., Expression 5.34 and related text) are here provided for the first time. Kempthorne's (1957) regression framework (Expressions 5.6–5.9 and related text) has been chosen as a common, convenient approach to develop all the models here provided—and also the additional results reserved for the next chapter—although some of the results had previously been obtained using other equivalent approaches.

The original NOIA model (Expressions 5.11 and 5.12, Álvarez-Castro and Carlborg 2007) and the two-dominance imprinting model (Expressions 5.29 and 5.30 above, Álvarez-Castro 2014) shown in Figs. 5.1 and 5.3, respectively, are two cases that were previously obtained with a different approach. In particular, they were obtained directly as regressions of the genotypic values on the gene content, as originally described by Fisher (1918) and shown graphically in those figures. The first step of this approach is to obtain the regression line, which in turn requires to obtain two parameters. One is the intercept, which is a point (shown as an empty circle in Figs. 5.1 and 5.3) given by the expectation of the gene content, $E(N) = p_{12}G_{12} + p_{12}G_{12} + 2p_{22}G_{22}$, and the population average, $E(G) = \mu$. The other parameter needed to obtain the regression line is its slope, which is the additive genetic effect, α, obtained as the covariance of the genotypic values and the gene content divided by the variance of the gene content, $cov(G,N)/var(N)$. To end up with, the dominance departures, δ_{ij}, are computed as the departures between the values provided by the regression line and the genotypic values. All the same, both this approach and the one followed along this chapter (using Kempthorne's (1957) setting) are equivalent and thus lead to an identical final result.

5.9.2 Matrix Notation and Orthogonality

In the end, while various methods may be used to obtain the desired quantitative genetics parameters providing the biological properties of the populations, our final aim has been to express the results in matrix notation (for the reasons discussed mainly in Chap. 3), both using the genetic-effects design matrix ($\mathbf{G} = \mathbf{SE}$) and its inverse ($\mathbf{E} = \mathbf{S}^{-1}\mathbf{G}$). The latest form expresses the genetic effects as a linear combination of the genotypic values. Whereas in the models at the individual level, addressed in the previous chapter, the genetic effects stood for effects of allele substitutions between individual genotypes and the genetic effects at the population level in the present chapter stand for effects of allele substitutions averaged over the genotypes present at a population. Thus, they reflect properties of the genetic system at a population, rather than of the genetic system itself.

As for the expression $\mathbf{G} = \mathbf{SE}$, it provides the genotypic values as a linear combination of the (additive, dominance...) genetic effects. This way, it uses the aforementioned properties of these genetic effects (in this chapter, at the population level) for partitioning both the genotypic values and the genetic variance into independent (i.e., orthogonal) components with a precise biological meaning. As orthogonality is a crucial fact of the theory provided in this chapter, we have revised and provided examples of what it means mathematically. We have also explained above some of the implications of orthogonality as a necessary mathematical property for the models to be biologically meaningful. In particular, we have shown that orthogonality allows for real partitioning of the genetic variance into a sum of variances alone—with no covariances between components implied—so that each term accounts for the contribution of one component of the model alone.

Nevertheless, it is at this point worth highlighting that the previous does not mean that orthogonality alone makes a model of genetic effects to be correct—i.e., it is not a sufficient condition for the model to be correct. In other words, the models not only need to be orthogonal but also ensure that the variables implied achieve the meaning we need them to have. Indeed, regressions analogous to Kempthorne's (1957) (Expressions 5.6, 5.14, 5.15, 5.21, 5.23, 5.24, 5.27, 5.28, 5.31, 5.34) have here been chosen amongst other ways to develop the formulae partly as a way for the readers to more easily track that meaning in the formulae.

5.9.3 An Evolving Quantitative Genetics

One important biological insight that can be extracted from the terms in the aforementioned decompositions is the response to selection on the phenotype in one generation step, as discussed in Chap. 2. In particular, the heritability $h^2 = V_A/V_P$ (Expression 2.25) has been used as the key quantitative genetics index in that regard, by expressing the response to selection as the heritability times the selection differential, $R = h^2\ S$ (Expression 2.28). In that chapter we have also pointed out, however, that strictly speaking such prediction is well grounded only under the assumption (amongst others) that the genetic architecture of the trait entails no significant epistasis or at least that in the population under study the terms V_{AA}, V_{AAA} ... are negligible. That assumption was necessary because classical quantitative genetics data are phenotypes and relatedness and it cannot be expected in general that acceptable estimates of all the variance components can be obtained from these data (see, e.g., Falconer and MacKay 1996).

Thus, classical quantitative genetics found ways to dodge most of what concerns the actual genetic architecture of the traits under study—which was for a long time completely out of reach, as discussed in the previous chapters. Now, for working with (more or less complex) explicit genetic architectures, classical quantitative genetics required to be significantly further implemented. The expressions provided in this chapter, as well as in the two adjacent chapters, are necessary to carry out such an endeavor, which makes particular sense as mapping experiments burst upon the scene. Indeed, properly detecting responsible loci and estimating the effects of their alleles in a sound, unbiased manner requires models general enough to accurately fit the data available, both in what regards the genotypic frequencies of the putative underlying loci (including all possible departures from equilibrium frequencies) and the genetic and gene-environment interactions to be tested. What is more, such models are required as well to analyze in depth the outcome of selection in the face of varying population facts and/or genetic architectures—more or less sustained by empirical observations.

Unfortunately, however, the aforementioned generality cannot be achieved by means of succinct mathematical developments. Thus, it becomes by no means surprising that this and the adjacent chapters of this book are profuse in formulae. Even then, these chapters are meant to be useful for readers not interested in checking every mathematical detail of each of the expressions provided. Indeed, the text also provides clarification on the issues found to be relevant for each of the cases covered. These include, for instance, clarifying the way Kempthorne's (1957) regressions must be solved in sequential steps (Expressions 5.6–5.10 and related text) or noticing the coincidence in the expressions of the female genetic effects of an X-linked locus when modeled with separate female and male effects (Expression 5.22) and those of an autosomal locus (Expression 5.11).

As mentioned above, the theory provided in this chapter shall be extended to departures from linkage equilibrium and to non-random associations between/among genes and environments. With this, a broad, unified theory of genetic effects—i.e., a general GP map accounting for environmental factors—shall be accomplished, both in what regards the individual and the population levels. Above and beyond, the next chapter also comprises a number of further extensions that make that GP more meaningful, which, incidentally, make the most of the matrix notation. Such extensions include additional genetic parameters, particularly Fisher's (1941) average excesses, a method for straightforwardly obtaining the decomposition of the genetic variance of a model of genetic effects at a population, and a convenient approach to estimate genetic effects from data that takes advantage of the models being orthogonal.

References

Álvarez-Castro JM (2014) Dissecting genetic effects with imprinting. Front Ecol Evol 2:51

Álvarez-Castro JM (2016) Genetic architecture. In: Wolf JB (ed) Encyclopedia of evolutionary biology. Oxford Academic Press, Oxford, pp 127–135

Álvarez-Castro JM, Carlborg Ö (2007) A unified model for functional and statistical epistasis and its application in quantitative trait loci analysis. Genetics 176:1151–1167

Álvarez-Castro JM, Crujeiras RM (2019) Orthogonal decomposition of the genetic variance for epistatic traits under linkage disequilibrium-applications to the analysis of bateson-dobzhansky-muller incompatibilities and sign epistasis. Front Genet 10:54

Álvarez-Castro JM, Yang R-C (2011) Multiallelic models of genetic effects and variance decomposition in non-equilibrium populations. Genetica 139:1119–1134

Álvarez-Castro JM, Carlborg O, Ronnegard L (2012) Estimation and interpretation of genetic effects with epistasis using the NOIA model. Methods Mol Biol 871:191–204

Avery OT, Macleod CM, Mccarty M (1944) Studies on the chemical nature of the substance inducing transformation of pneumococcal types nducton of transformation by a desoxyribonucleic acid fraction isolated from pneumococcus type ii. J Exp Med 79:137–158

Cheverud JM, Routman EJ (1995) Epistasis and its contribution to genetic variance components. Genetics 139:1455–1461

Cockerham CC (1954) An extension of the concept of partitioning hereditary variance for analysis of covariances among relatives when epistasis is present. Genetics 39:859–882

Deng S, Hardin J, Amos CI, Xiao F (2020) Joint modeling of eQTLs and parent-of-origin effects using an orthogonal framework with RNA-seq data. Hum Genet 139:1107–1117

Draper NR, Smith H (1998) Applied Regression Analysis. New York, John Wiley & Sons

Falconer DS, Mackay TFC (1996) Introduction to quantitative genetics. Prentice Hall, Harlow

Fisher RA (1918) The correlation between relatives on the supposition of Mendelian inheritance. Trans Roy Soc Edinburgh 52:339–433

Fisher RA (1930) The genetical theory of natural selection. Clarendon, Oxford

Fisher RA (1941) Average excess and average effect of a gene substitution. Ann Eugen 11:53–63

Franklin R, Gosling RG (1953) Molecular configuration in sodium thymonucleate. Nature 171: 740–741

Hansen TF, Wagner GP (2001) Modeling genetic architecture: a multilinear theory of gene interaction. Theor Popul Biol 59:61–86

Harville DA (1997) Matrix algebra from a statistician's perspective. Springer, New York

Hershey AD, Chase M (1952) Independent functions of viral protein and nucleic acid in growth of bacteriophage. J Gen Physiol 36:39–56

Kempthorne O (1954) The correlation between relatives in a random mating population. Proc R Soc Lond B Biol Sci 143:102–113

Kempthorne O (1957) An introduction to genetic statistics. Wiley, New York

Kempthorne O (1968) The correlation between relatives on the supposition of Mendelian inheritance. Am J Hum Genet 20:402–403

Lander ES, Botstein D (1989) Mapping mendelian factors underlying quantitative traits using RFLP linkage maps. Genetics 121:185–199

Ma J, Xiao F, Xiong M, Andrew AS, Brenner H, Duell EJ, Haugen A, Hoggart C, Hung RJ, Lazarus P, Liu C, Matsuo K, Mayordomo JI, Schwartz AG, Staratschek-Jox A, Wichmann E, Yang P, Amos CI (2012) Natural and orthogonal interaction framework for modeling gene-environment interactions with application to lung cancer. Hum Hered 73:185–194

Morgan TH (1910) Sex limited inheritance in drosophila. Science 32:120–122

Provine WB (1971) The origins of theoretical population genetics. University of Chicago Press, Chicago

Provine WB (1986) Sewall Wright and evolutionary biology. University of Chicago Press, Chicago

Stoltzfus A, Cable K (2014) Mendelian-mutationism: the forgotten evolutionary synthesis. J Hist Biol 47:501–546

Tiwari HK, Elston RC (1997) Deriving components of genetic variance for multilocus models. Genet Epidemiol 14:1131–1136

Van Loan C (2000) The ubiquitous Kronecker products. J Comput Appl Math 123:85–100

Watson JD, Crick FHC (1953) A structure for deoxyribose nucleic acid. Nature 171:737–738

Weinreich DM, Watson RA, Chao L (2005) Perspective: Sign epistasis and genetic constraint on evolutionary trajectories. Evolution 59:1165–1174

Yang R-C (2004) Epistasis of quantitative trait loci under different gene action models. Genetics 167:1493–1505

Yang R-C, Álvarez-Castro JM (2008) Functional and statistical genetic effects with miltiple alleles. Curr Topics Genet 3:49–62

Zeng ZB, Wang T, Zou W (2005) Modeling quantitative trait Loci and interpretation of models. Genetics 169:1711–1725

Check for updates

A General Theory of Genetic Effects 6

Abstract

The natural and orthogonal interactions (NOIA) model was originally developed not only as one more step in the improvement of genetic models but actually as a sound basis for the establishment of a completely general and unified theory of genetic effects—i.e., a general genotype-to-phenotype (GP) map. In brief, that theory provides decompositions of the genotypic values into genetic effects, both at the individual and at the population levels. Both decompositions must account for all kinds of peculiarities of complex genetic architectures (by modeling them mathematically), and the latter must also properly fit any departures of the populations from equilibrium frequencies. We here expound a procedure that provides an orthogonal decomposition of the genotypic values accounting for non-random associations between/among the factors (genes and environments) involved—the correlation-wise orthogonal interactions (COIA) regression framework. In particular, this chapter gathers both already published and novel developments to satisfactorily deal with such non-random associations, and implement them into an associations-resolved NOIA (ARNOIA)—an upgrade of NOIA to a 2.0 version. This theory, coupled with the developments presented in the previous two chapters, comprises a major advance in the generalization of the theory of genetic effects.

6.1 Introduction

The natural and orthogonal interactions (NOIA) model was originally developed to set the basis for a general genotype-to-phenotype (GP) map (Álvarez-Castro and Carlborg 2007). NOIA's first installment already merged several advances that had been considered separately in previous publications, by providing a multilocus two-allele setting accounting for arbitrary epistasis and departures from the Hardy-Weinberg proportions and integrating that theory together with a detailed description

J. M. Álvarez-Castro, *Genes, Environments and Interactions*,
https://doi.org/10.1007/978-3-031-41159-5_6

in terms of allele substitutions between individual genotypes. The labels "natural" and "orthogonal" stand for what we have referred to in previous chapters as the individual and the population levels, respectively—which NOIA for the first time unified under a common, versatile mathematical framework. It is key to keep in mind that these levels of analysis entail two complementary ways of looking at the same observable fact—the underlying genetics of a trait.

Particularly when dealing with epistasis, models aiming to reflect properties of allele substitutions without considering the population level were valued. For instance, Jana (1972) found the "simpler and more direct" models by Seyffert (1966) to be preferable to the population-involved "sophisticated biometrical techniques [. . .] when the aim of an investigation is to seek for a physiological and biochemical interpretation of dominance and epistasis" (see also Giesel 1977; Álvarez-Castro 2012). Later on, with gene mapping in mind, this kind of approaches to the study of epistasis has been referred to as physiological (Cheverud and Routman 1995), functional (Hansen and Wagner 2001), genetic, biological (Moore 2005; Moore and Williams 2005), and compositional (Phillips 2008).

Although all these approaches try to keep away from the population level, throughout this book we shall consider as models at the individual level only those with individual genotypes as reference points, for the reasons stated in Chap. 4. We do so despite the fact that the original publication of NOIA considered a wider scope of non-population models (Álvarez-Castro and Carlborg 2007). In subsequent publications, NOIA was extended to additional genetic architectures at the individual level, like multiple alleles (Yang and Álvarez-Castro 2008), gene-environment interactions (Ma et al. 2012), and imprinting (Álvarez-Castro 2014), thus being upgraded to a more general genotype-to-phenotype (GP) map accounting also for environmental factors.

Concerning now formulations aiming to match the population level, they were often referred to—once more, particularly when dealing with epistasis—as statistical (e.g., again, Cheverud and Routman 1995; Hansen and Wagner 2001; Moore 2005; Moore and Williams 2005; Phillips 2008). These formulations aim to analyze the same phenomenon as the aforementioned ones (epistasis), although focusing on the implications of it at the population level. For addressing this task properly, their parameters have to be averages over the population of the genetic effects at the individual level, which amongst other things implies that they must fulfill a statistical property called orthogonality—explained in detail in the previous chapter. In this regard, the extensions of NOIA need not only to consider increasingly complex genetic architectures, but also all kinds of departures of the populations from equilibrium frequencies—in order to orthogonally fit them (e.g. Álvarez-Castro and Carlborg 2007; Álvarez-Castro and Yang 2011; Álvarez-Castro 2014, 2020; Álvarez-Castro and Crujeiras 2019).

As mentioned above, the original publication of NOIA properly implemented departures from the Hardy-Weinberg proportions. Such advantage was kept in posterior extensions of NOIA at the population level to multiple alleles (Álvarez-Castro and Yang 2011), gene-environment interactions (Ma et al. 2012), and imprinting (Álvarez-Castro 2014). However, those extensions assumed random associations between/among factors—i.e., linkage equilibrium and random associations between genes and environments, whenever several loci and gene-environment interactions were considered, respectively. In this chapter, we shall provide the extension of NOIA, at the population level, to arbitrary departures from linkage equilibrium frequencies—thus including linkage disequilibrium—and to arbitrary, non-random associations between genes and environments, even beyond the extensions to these conditions attained by Álvarez-Castro and Crujeiras (2019) and Álvarez-Castro (2020), respectively.

These extensions are crucial from the biological standpoint, to which the mathematical standpoint sheds light. Originally, NOIA was built (as discussed in the previous chapter) on the basis of the matrix notation as established by Tiwari and Elston (1997) and tuned at the population level by Zeng et al. (2005)—explained in Chap. 3. In brief, following Tiwari and Elston (1997), single-locus genetic-effects design matrices are developed and combined (using the Kronecker product) for obtaining a multilocus model. The single-locus matrices do not only fit properly (therefore, orthogonally) allele frequencies of the population, as implemented by Zeng et al. (2005), but also single-locus genotype frequencies (Álvarez-Castro and Carlborg 2007; Álvarez-Castro and Yang 2011; Álvarez-Castro 2014).

That scheme for obtaining a multilocus genetic-effects design matrix as a Kronecker product of single-locus settings is however limited to linkage equilibrium—and to random associations between genes and environments whenever gene-environment interactions are considered. Thus, in order to overcome such limitations, multilocus genetic-effects design matrices have to be developed directly (Álvarez-Castro and Crujeiras 2019; Álvarez-Castro 2020)—a procedure known as the COIA regression framework (Álvarez-Castro 2020). In this chapter, we present the COIA regression framework and implement it into the NOIA model. With this, NOIA gets upgraded to arbitrary associations between/among factors (loci and environments)—a 2.0 version of NOIA that has been coined as associations-resolved NOIA (ARNOIA). With that, a fairly general theory of genetic effects—a general GP map—is provided.

6.2 The COIA Regression Framework

The matrix notation of models of genetic effects enables the development genetic-effects design matrices of multilocus models as Kronecker products of single-locus genetic-effects design matrices (e.g., Tiwari and Elston 1997; Zeng et al. 2005; Álvarez-Castro and Carlborg 2007; see Chap. 3). This way, the scalars of the resulting multilocus matrices are products of scalars of the original single-locus matrices, and thus, they imply products of marginal frequencies. On the other hand, the multilocus genotypic frequencies under linkage equilibrium are actually products of the marginal genotypic frequencies—e.g., for two loci, A^1 and A^2, $p_{ijkl}=p^1_{ij}p^2_{kl}$. This is to say, it looks sensible that the scope of the multilocus models obtained as Kronecker products of the single-locus models is restricted to linkage equilibrium.

When linkage disequilibrium occurs due to physical linkage, the multilocus frequencies can be determined by one additional parameter, the coefficient of linkage disequilibrium, or disequilibrium index, D. Indeed, this index was designed to measure non-random associations between alleles that persist due to the decay of recombination between neighboring loci—i.e., due to physical linkage (see Lewontin and Kojima 1960). However, departures from linkage equilibrium can also affect loci in different linkage groups, e.g., due to selection (particularly when combined with epistasis), non-random mating, drift, and gene flow. In order to account for that more general situation, linkage disequilibrium turned to be more frequently called gametic phase disequilibrium instead—a name not yet absent of criticism, as linkage disequilibrium may not even imply lack of equilibrium (e.g., Slatkin 2008).

In any case, it is noteworthy that the theory provided below accounts for completely arbitrary multilocus genotypic frequencies, thus beyond not only those under linkage equilibrium, but further beyond those that can be obtained as departures from linkage equilibrium measured by the coefficient of linkage disequilibrium, D (i.e., beyond those under linkage disequilibrium as originally conceived). Indeed, for determining, for instance, completely arbitrary genotypic frequencies of two biallelic loci, eight values are required—since the frequency of the ninth genotype can be obtained as one minus the sum of the frequencies of the other eight. Hence, they cannot be determined by the marginal frequencies plus the coefficient of linkage disequilibrium—summing up to a maximum of five independent values.

We shall actually begin by obtaining the pertinent quantitative genetic parameters for two biallelic autosomal loci with arbitrary genotypic frequencies, and then we shall show how it can be extended to more complex situations. Afterwards, we shall address non-random associations between genes and environments. Overall, the method below provides an extension of the theory presented in the previous chapter beyond linkage equilibrium and random-associations between genes and environments. In other words, it enables us to attain appropriate, orthogonal decompositions of the genotypic values and of the genetic variance in the face of correlations between/among factors (both genes and loci) occur. This is the reason why these developments are labeled as the correlation-wise orthogonal interactions (COIA) regression framework (Álvarez-Castro 2020).

6.2.1 Two Biallelic Autosomal Loci

Let us start by recalling the regression of the case of one biallelic locus A (Expression 5.6 and related text in Chap. 5), which decomposes the genotypic values into additive and dominance components from the reference of the population average phenotype as $\mathbf{G} = \mathbf{1}\,\mu + \mathbf{N}\boldsymbol{\alpha} + \boldsymbol{\delta}$, expanding to:

$$\begin{pmatrix} G_{11} \\ G_{12} \\ G_{22} \end{pmatrix} = \begin{pmatrix} 1 \\ 1 \\ 1 \end{pmatrix} \mu + \begin{pmatrix} 2 & 0 \\ 1 & 1 \\ 0 & 2 \end{pmatrix} \begin{pmatrix} \alpha_1 \\ \alpha_2 \end{pmatrix} + \begin{pmatrix} \delta_{11} \\ \delta_{12} \\ \delta_{22} \end{pmatrix}, \tag{6.1}$$

With this regression in mind, let us now consider two autosomal biallelic loci, A^1 and A^2, with completely arbitrary genotypic frequencies, p_{ijkl}, ij, kl = 11, 12, 22. Applying a methodology analogous to that used throughout the previous chapter, the genotypic values of such genetic system, $\mathbf{G} = (G_{ijkl})$, could be partitioned into additive, dominance, and epistasis components using a regression framework of the type:

$$\mathbf{G} = \mathbf{N}_\mu \mu + \mathbf{N}_\alpha \boldsymbol{\alpha} + \mathbf{N}_\delta \boldsymbol{\delta} + \boldsymbol{\eta}, \tag{6.2}$$

where the mean phenotype, $\mu = \Sigma p_{ijkl} G_{ijkl}$, is the first explanatory variable; $\boldsymbol{\alpha}$ and $\boldsymbol{\delta}$ are the other two explanatory variables; the design matrix of the mean phenotype is $\mathbf{N}_\mu = \mathbf{1}^{(9)}$ (a column vector of nine ones); $\mathbf{N}_\alpha$ and $\mathbf{N}_\delta$ are the other two design matrices; and the error term, $\boldsymbol{\eta}$, is the epistatic component.

However, as explained in Chap. 3, the genetic-effects design matrix of a two-locus genetic system implies to further decompose the epistatic component into additive-by-additive, dominance-by-additive, additive-by-dominance, and dominance-by-dominance components. Thus, a complete decomposition of the genetic system may be attained by further developing Expression 6.2 into a regression framework of the type:

$$\mathbf{G} = \mathbf{N}_\mu \mu + \mathbf{N}_\alpha \boldsymbol{\alpha} + \mathbf{N}_\delta \boldsymbol{\delta} + \mathbf{N}_{\alpha\alpha} \boldsymbol{\alpha\alpha} + (\mathbf{N}_{\delta\alpha} | \mathbf{N}_{\alpha\delta})(\boldsymbol{\delta\alpha} | \boldsymbol{\alpha\delta})^{\mathrm{T}} + \boldsymbol{\delta\delta}, \tag{6.3}$$

where additive-by-dominance and dominance-by-additive, as equal level components, are merged into a single term (Álvarez-Castro and Crujeiras 2019).

As also explained throughout the previous chapter, the explanatory variables in Expression 6.3 must not be obtained altogether (see particularly Expression 5.10 and related text in Chap. 5). They must be obtained using a sequential procedure instead, which we shall hereafter describe following Álvarez-Castro (2020; see Expression 3 in that publication). Let $\mathbf{I}^{(9)}$ be the identity matrix of dimension nine and let $\mathbf{P}$ be the weights matrix—i.e., the diagonal matrix of the genotypic frequencies, $\mathbf{P} = diag(p_{ijkl})$. Let us also number the design matrices and the explanatory variables as $\mathbf{N}_1 = \mathbf{N}_\alpha$, $\boldsymbol{\upsilon}_1 = \boldsymbol{\alpha}$, $\mathbf{N}_2 = \mathbf{N}_{\alpha\delta}$, $\boldsymbol{\upsilon}_2 = \boldsymbol{\delta}$, $\mathbf{N}_3 = \mathbf{N}_{\alpha\alpha}$, $\boldsymbol{\upsilon}_3 = \boldsymbol{\alpha\alpha}$, $\mathbf{N}_4 = (\mathbf{N}_{\delta\alpha} | \mathbf{N}_{\alpha\delta})$, and $\boldsymbol{\upsilon}_4 = (\boldsymbol{\delta\alpha} | \boldsymbol{\alpha\delta})^{\mathrm{T}}$, and let $\boldsymbol{\eta}_0$ be the mean-corrected vector of genotypic values, i.e., $\boldsymbol{\eta}_0 = \mathbf{G} - \mathbf{N}_\mu \mu$. Then, the solutions to Expression 6.3, $\boldsymbol{\upsilon}_c$, $c = 1, \ldots, 4$, can typically be obtained recurrently as:

$$\boldsymbol{\upsilon}_c = \tilde{\mathbf{H}}_c \boldsymbol{\eta}_{c-1} \text{ and } \boldsymbol{\eta}_c = \mathbf{M}_c \boldsymbol{\eta}_{c-1}, c = 1, \ldots, 4, \tag{6.4}$$

where $\tilde{\mathbf{H}}_c = \left(\mathbf{N}_c^{\mathrm{T}} \mathbf{P} \mathbf{N}_c\right)^{-1} \mathbf{N}_c^{\mathrm{T}} \mathbf{P}$ and $\mathbf{M}_c = \mathbf{I}^{(9)} - \mathbf{N}_c \tilde{\mathbf{H}}_c$.

There are two important remarks to make for the previous to provide the desired results in practice. In the first place, we still need to develop explicitly the design matrices of the genetic system, $\mathbf{N}_c$, $c = 1,\ldots, 4$. Second, given those design matrices, the standard procedure described right above to obtain the matrices $\tilde{\mathbf{H}}_c$, $c = 1,\ldots, 4$, (the one used throughout the previous chapter and by Álvarez-Castro (2020), i.e., $\tilde{\mathbf{H}}_c = \left(\mathbf{N}_c^{\mathrm{T}}\mathbf{P}\mathbf{N}_c\right)^{-1}\mathbf{N}_c^{\mathrm{T}}\mathbf{P}$) may not work for all steps, in which case an alternative procedure has to be used instead (Álvarez-Castro and Crujeiras 2019). Both remarks are addressed hereafter, focusing on the components 1 to 4 (additive to additive-by-dominance) one by one.

Concerning first the additive component, the design matrix $\mathbf{N}_1 = \mathbf{N}_\alpha$ can be obtained by just extending the logic followed at the one-locus case (see Expression 6.1) to two loci. In particular, we could expand a truncated (after the additive component) part of Expression 6.2, $\mathbf{G} = \mathbf{N}_\mu\mu + \mathbf{N}_\alpha\boldsymbol{\alpha} + \boldsymbol{\eta}_\alpha$, alternatively $\mathbf{G} = \mathbf{N}_\mu\mu + \mathbf{N}_1\boldsymbol{\upsilon}_1 + \boldsymbol{\eta}_1$, as (Álvarez-Castro and Crujeiras 2019):

$$\begin{pmatrix} G_{1111} \\ G_{1211} \\ G_{2211} \\ G_{1112} \\ G_{1212} \\ G_{2212} \\ G_{1122} \\ G_{1222} \\ G_{2222} \end{pmatrix} = \begin{pmatrix} 1 \\ 1 \\ 1 \\ 1 \\ 1 \\ 1 \\ 1 \\ 1 \\ 1 \end{pmatrix}\mu + \begin{pmatrix} 2 & 0 & 2 & 0 \\ 1 & 1 & 2 & 0 \\ 0 & 2 & 2 & 0 \\ 2 & 0 & 1 & 1 \\ 1 & 1 & 1 & 1 \\ 0 & 2 & 1 & 1 \\ 2 & 0 & 0 & 2 \\ 1 & 1 & 0 & 2 \\ 0 & 2 & 0 & 2 \end{pmatrix} \begin{pmatrix} \alpha_1^1 \\ \alpha_2^1 \\ \alpha_1^2 \\ \alpha_2^2 \end{pmatrix} + \begin{pmatrix} \eta_{1111}^\alpha \\ \eta_{1211}^\alpha \\ \eta_{2211}^\alpha \\ \eta_{1112}^\alpha \\ \eta_{1212}^\alpha \\ \eta_{2212}^\alpha \\ \eta_{1122}^\alpha \\ \eta_{1222}^\alpha \\ \eta_{2222}^\alpha \end{pmatrix}, \tag{6.5}$$

where the superscripts of the additive parameters are locus indicators and the subscripts are allele indicators. Note that the rows of $\mathbf{N}_\alpha$ in this expression come simply from combining the rows of the corresponding matrix of the one-locus case, $\mathbf{N}$ (see Expression 6.1).

Now, the additive parameters, α_j^i, $i, j = 1, 2$, would typically be obtained as $\boldsymbol{\alpha} = \boldsymbol{\upsilon}_1 = \tilde{\mathbf{H}}_1\boldsymbol{\eta}_0$, with $\tilde{\mathbf{H}}_1 = \left(\mathbf{N}_1^{\mathrm{T}}\mathbf{P}\mathbf{N}_1\right)^{-1}\mathbf{N}_1^{\mathrm{T}}\mathbf{P}$, as shown in Expression 6.4, particularly assuming $c = 1$. However, the determinant of matrix $\mathbf{N}_1^{\mathrm{T}}\mathbf{P}\mathbf{N}_1$ equals zero (i.e., the matrix is singular) and, hence, it is not possible to compute its inverse, $\left(\mathbf{N}_1^{\mathrm{T}}\mathbf{P}\mathbf{N}_1\right)^{-1}$, as required in Expression 6.4. Therefore, as advanced above, an alternative method has to be used to obtain $\tilde{\mathbf{H}}_1$. That alternative method is described hereafter for the general case (modified from Álvarez-Castro and Crujeiras 2019):

First, the eigenvalues and eigenvectors of singular matrix $\mathbf{N}_c^{\mathrm{T}}\mathbf{P}\mathbf{N}_c$ (with $c = 1$ in our current case) must be computed—in practice, this can be done, for instance, using the appropriate built-in commands of R Core Team (2017). Next, a diagonal matrix, $\mathbf{D}_c$ (in our current case, $\mathbf{D}_1$, i.e., $\mathbf{D}_\alpha$), is built with the non-nil eigenvalues so obtained, while their corresponding eigenvectors become the columns of matrix $\mathbf{U}_c$ (in our current case, $\mathbf{U}_1$, i.e., $\mathbf{U}_\alpha$). Then, the solution, $\boldsymbol{\upsilon}_c$ (in our current case, $\boldsymbol{\upsilon}_1$, i.e., $\boldsymbol{\alpha}$), may be obtained as indicated in Expression 6.4, by just replacing, in the singular matrix ($\mathbf{N}_c^{\mathrm{T}}\mathbf{P}\mathbf{N}_c$), the design matrix ($\mathbf{N}_c$) and the weight matrix ($\mathbf{P}$) by the transpose and the inverse of the matrices obtained just above, i.e., $\mathbf{U}_c^{\mathrm{T}}$ and $\mathbf{D}_c^{-1}$, respectively. Thus in general terms, the alternative method may be expressed as:

$$\boldsymbol{\upsilon}_c = \tilde{\mathbf{H}}_c \boldsymbol{\eta}_{c-1}, \text{ with } \tilde{\mathbf{H}}_c = \left(\mathbf{U}_c \mathbf{D}_c^{-1} \mathbf{U}_c^{\mathrm{T}}\right)^{-1} \mathbf{N}_c^{\mathrm{T}} \mathbf{P}. \tag{6.6}$$

In our current case, we obtain $\boldsymbol{\alpha} = \boldsymbol{\upsilon}_1$ by applying this general expression for the particular case $c = 1$.

Second, we move on to the dominance component. In the regression of the one-locus case (Expression 6.1), dominance is the error term and, hence, it has no design matrix. We can thus interpret that the matrix multiplying that term is an identity matrix with the adequate dimension, which is three. It is therefore not surprising that, analogous to the additive component above (Expression 6.5), the design matrix of the dominance component in the two-locus case can be built with combinations of rows of an identity matrix of dimension three, $\mathbf{I}^{(3)}$. The regression step to obtain the dominance component can thus be expressed as $\boldsymbol{\eta}_\alpha = \mathbf{N}_\delta \boldsymbol{\delta} + \boldsymbol{\eta}_\delta$, alternatively $\boldsymbol{\eta}_1 = \mathbf{N}_2 \boldsymbol{\upsilon}_2 + \boldsymbol{\eta}_2$, expanding to Álvarez-Castro and Crujeiras (2019):

$$\begin{pmatrix} \eta^\alpha_{1111} \\ \eta^\alpha_{1211} \\ \eta^\alpha_{2211} \\ \eta^\alpha_{1112} \\ \eta^\alpha_{1212} \\ \eta^\alpha_{2212} \\ \eta^\alpha_{1122} \\ \eta^\alpha_{1222} \\ \eta^\alpha_{2222} \end{pmatrix} = \begin{pmatrix} 1 & 0 & 0 & 1 & 0 & 0 \\ 0 & 1 & 0 & 1 & 0 & 0 \\ 0 & 0 & 1 & 1 & 0 & 0 \\ 1 & 0 & 0 & 0 & 1 & 0 \\ 0 & 1 & 0 & 0 & 1 & 0 \\ 0 & 0 & 1 & 0 & 1 & 0 \\ 1 & 0 & 0 & 0 & 0 & 1 \\ 0 & 1 & 0 & 0 & 0 & 1 \\ 0 & 0 & 1 & 0 & 0 & 1 \end{pmatrix} \begin{pmatrix} \delta^1_{11} \\ \delta^1_{12} \\ \delta^1_{22} \\ \delta^2_{11} \\ \delta^2_{12} \\ \delta^2_{22} \end{pmatrix} + \begin{pmatrix} \eta^\delta_{1111} \\ \eta^\delta_{1211} \\ \eta^\delta_{2211} \\ \eta^\delta_{1112} \\ \eta^\delta_{1212} \\ \eta^\delta_{2212} \\ \eta^\delta_{1122} \\ \eta^\delta_{1222} \\ \eta^\delta_{2222} \end{pmatrix}. \tag{6.7}$$

Also as in the additive component above, the solution to this regression has to be obtained using Expression 6.4, modified by the alternative procedure described in Expression 6.6, with the only difference that now $c = 2$.

Third, we address the component of the additive-by-additive epistatic interaction. As shown by Álvarez-Castro and Crujeiras (2019), for the two-locus case, the design matrix of this component can be obtained as the Kronecker product of the matrices of the single-locus case (Expression 6.1). Using the Kronecker product of genetic-effects design, matrices at the population level (in Chaps. 3 and 5) implied that the resulting model was limited to linkage equilibrium. Indeed, these single-locus matrices were built with single-locus genotypic frequencies. Here, we still use the Kronecker product, but earlier in the construction of the model—we use it, specifically, when building the two-locus design matrices of the regression framework. These two-locus design matrices enable us to also implement a two-locus weights matrix, **P** (Expressions 6.4 and 6.6), with arbitrary (thus, beyond linkage equilibrium) two-locus frequencies. The regression step to obtain the additive-by-additive component can be expressed as $\boldsymbol{\eta}_\delta = \mathbf{N}_{\alpha\alpha}\boldsymbol{\alpha\alpha} + \boldsymbol{\eta}_{\alpha\alpha}$, alternatively $\boldsymbol{\eta}_2 = \mathbf{N}_3\boldsymbol{\upsilon}_3 + \boldsymbol{\eta}_3$, expanding to:

$$\begin{pmatrix} \eta^{\delta}_{1111} \\ \eta^{\delta}_{1211} \\ \eta^{\delta}_{2211} \\ \eta^{\delta}_{1112} \\ \eta^{\delta}_{1212} \\ \eta^{\delta}_{2212} \\ \eta^{\delta}_{1122} \\ \eta^{\delta}_{1222} \\ \eta^{\delta}_{2222} \end{pmatrix} = \begin{pmatrix} 4 & 0 & 0 & 0 \\ 2 & 2 & 0 & 0 \\ 0 & 4 & 0 & 0 \\ 2 & 0 & 2 & 0 \\ 1 & 1 & 1 & 1 \\ 0 & 2 & 0 & 2 \\ 0 & 0 & 4 & 0 \\ 0 & 0 & 2 & 2 \\ 0 & 0 & 0 & 4 \end{pmatrix} \begin{pmatrix} \alpha\alpha_{11} \\ \alpha\alpha_{12} \\ \alpha\alpha_{21} \\ \alpha\alpha_{22} \end{pmatrix} + \begin{pmatrix} \eta^{\alpha\alpha}_{1111} \\ \eta^{\alpha\alpha}_{1211} \\ \eta^{\alpha\alpha}_{2211} \\ \eta^{\alpha\alpha}_{1112} \\ \eta^{\alpha\alpha}_{1212} \\ \eta^{\alpha\alpha}_{2212} \\ \eta^{\alpha\alpha}_{1122} \\ \eta^{\alpha\alpha}_{1222} \\ \eta^{\alpha\alpha}_{2222} \end{pmatrix}. \tag{6.8}$$

In this case, the solution can be obtained using the conventional method (Álvarez-Castro and Crujeiras 2019), i.e., applying Expression 6.4, with $c = 3$.

Last, we solve the combined component entailing additive-by-dominance and dominance-by-additive epistatic interactions (following once again Álvarez-Castro and Crujeiras 2019). Analogous to the additive-by-additive component right above, the design matrix of the regression can be obtained by means of a Kronecker product. To be precise, we now need two matrices, one for the additive-by-dominance and another one for the dominance-by-additive components (which we shall afterwards combine into one single matrix) and, hence, two Kronecker products.

The matrix of the dominance-by-additive component, $\mathbf{N}_{\delta\alpha}$, can be obtained as the Kronecker product of the identity matrix of dimension 3—a matrix already used when developing the dominance component above (Expression 6.7 and related text)—and the design matrix of the single-locus additive component. The matrix of the additive-by-dominance component, $\mathbf{N}_{\alpha\delta}$, can be obtained as the Kronecker product of the same matrices in the reverse order. As noted in Chap. 3, the factors of the Kronecker products are displayed in the reverse order, i.e., $\mathbf{N}_{\delta\alpha} = \mathbf{N}^1_\alpha \otimes \mathbf{I}^{(3)}$, and $\mathbf{N}_{\alpha\delta} = \mathbf{I}^{(3)} \otimes \mathbf{N}^2_\alpha$ where $\mathbf{N}^1_\alpha$ and $\mathbf{N}^2_\alpha$ are the design matrices of the one-locus additive component for loci A^1 and A^2, respectively, which are actually equal and also equal to **N** in Expression 6.1. Then the desired design matrix results from just concatenating the two matrices as $\mathbf{N}_4 = (\mathbf{N}_{\delta\alpha}|\mathbf{N}_{\alpha\delta})$. This leads to the last regression step $\boldsymbol{\eta}_{\alpha\alpha} = (\mathbf{N}_{\delta\alpha}|\mathbf{N}_{\alpha\delta})(\boldsymbol{\delta\alpha}|\boldsymbol{\alpha\delta})^{\mathrm{T}} + \boldsymbol{\varepsilon\delta\delta}$, alternatively $\boldsymbol{\eta}_3 = \mathbf{N}_4\boldsymbol{\upsilon}_4 + \boldsymbol{\eta}_4$, expanding to:

$$\begin{pmatrix} \eta^{\alpha\alpha}_{1111} \\ \eta^{\alpha\alpha}_{1211} \\ \eta^{\alpha\alpha}_{2211} \\ \eta^{\alpha\alpha}_{1112} \\ \eta^{\alpha\alpha}_{1212} \\ \eta^{\alpha\alpha}_{2212} \\ \eta^{\alpha\alpha}_{1122} \\ \eta^{\alpha\alpha}_{1222} \\ \eta^{\alpha\alpha}_{2222} \end{pmatrix} = \left(\begin{array}{cccccc|cccccc} 2 & 0 & 0 & 0 & 0 & 0 & 2 & 0 & 0 & 0 & 0 & 0 \\ 0 & 2 & 0 & 0 & 0 & 0 & 1 & 1 & 0 & 0 & 0 & 0 \\ 0 & 0 & 2 & 0 & 0 & 0 & 0 & 2 & 0 & 0 & 0 & 0 \\ 1 & 0 & 0 & 1 & 0 & 0 & 0 & 0 & 2 & 0 & 0 & 0 \\ 0 & 1 & 0 & 0 & 1 & 0 & 0 & 0 & 1 & 1 & 0 & 0 \\ 0 & 0 & 1 & 0 & 0 & 1 & 0 & 0 & 0 & 2 & 0 & 0 \\ 0 & 0 & 0 & 2 & 0 & 0 & 0 & 0 & 0 & 0 & 2 & 0 \\ 0 & 0 & 0 & 0 & 2 & 0 & 0 & 0 & 0 & 0 & 1 & 1 \\ 0 & 0 & 0 & 0 & 0 & 2 & 0 & 0 & 0 & 0 & 0 & 2 \end{array}\right) \left(\begin{array}{c|c} \delta\alpha_{111} & \alpha\delta_{111} \\ \delta\alpha_{121} & \alpha\delta_{211} \\ \delta\alpha_{221} & \alpha\delta_{112} \\ \delta\alpha_{112} & \alpha\delta_{212} \\ \delta\alpha_{122} & \alpha\delta_{122} \\ \delta\alpha_{222} & \alpha\delta_{222} \end{array}\right)^{\mathrm{T}} + \begin{pmatrix} \delta\delta_{1111} \\ \delta\delta_{1211} \\ \delta\delta_{2211} \\ \delta\delta_{1112} \\ \delta\delta_{1212} \\ \delta\delta_{2212} \\ \delta\delta_{1122} \\ \delta\delta_{1222} \\ \delta\delta_{2222} \end{pmatrix}, \tag{6.9}$$

where the transpose operation is meant to affect only the relative position of the two blocks of the array of the explanatory variable $(\boldsymbol{\delta\alpha}|\boldsymbol{\alpha\delta})^{\mathrm{T}}$, hence leading to a column vector.

As it is the case for the additive and for the dominance components above, the solution of Expression 6.9 can be obtained using Expression 6.4 modified by the alternative procedure described in Expression 6.6, assuming particularly $c = 4$. As pointed out by Álvarez-Castro and Crujeiras (2019), it is noteworthy that this last step (Expression 6.9) can just as well be further split into two sequential regression steps—one for the additive-by-dominance and another one for the dominance-by-additive interaction components. If this latest option is chosen, then the solutions to each of the two regressions can be obtained with just the standard procedure in Expression 6.4, i.e., with no need of the alternative procedure in Expression 6.6, which is reassuring about the validity of this alternative procedure. Indeed, obtaining the same results with an option that bypasses the alternative procedure in Expression 6.6 provides a successful verification on the validity of this alternative procedure.

6.2.2 Increasingly Complex Genetic Architectures

For the case of two biallelic autosomal loci above, all required design matrices have been obtained from the single-locus design matrices, either by combining their rows (for the components accounting for effects of the marginal loci) or by means of a Kronecker product (for the components accounting for effects of interactions between the two loci). The procedure works in an analogous way when the single-locus matrices are different from those of biallelic loci. Single-locus matrices of a wide range of cases have been provided in Chap. 5. For instance, if one of the loci has three alleles, the design matrix of the single-locus additive component can be taken from Expression 5.14, and the design matrix of the dominance component is the identity matrix of dimension six.

The procedure developed above for a two-locus case holds in general terms also in case more loci are considered. Regression in Expression 6.2 can actually be applied as is to an arbitrary number of loci. It would thus only be necessary to develop larger design matrices accounting for the appropriate number of loci (with combinations of rows of more matrices and also larger Kronecker products). Although Expression 6.3 would not keep its exact form with more loci, it can just be extended to account for the emerging interaction terms arising with each new locus considered. Such emerging interaction terms have been introduced in Chap. 3.

6.2.3 One Biallelic Locus and Two Environmental Conditions

As we have gone through above, the gene-gene interplay at the population level involves not only whether there is interaction in what regards the phenotype each genotype is expected to lead to (i.e., a genetic architecture with epistasis) but also whether there are departures from random associations between the frequencies of the marginal genotypes at the level of the population—i.e., departures from the linkage equilibrium frequencies. Analogously, a gene and an environment affecting a trait may display gene-environment interaction at the level of the genetic architecture (as dealt with in the previous chapter), and it may also display departures from random associations between the frequencies of the genotypes and those of the adscription to the environmental variants—i.e., gene-environment correlation.

Hereafter, we shall present (modified from Álvarez-Castro 2020) the theoretical developments adequate to deal with gene-environment interaction and gene-environment correlation, simultaneously—i.e., adequate to address a general gene-environment interplay. These developments are thus beyond the population facts addressed in the previous chapter, where several multialleic loci with arbitrary departures from the Hardy-Weinberg proportions and several environmental variables under arbitrary numbers of conditions were already considered.

We initially consider a biallelic locus A (with alleles A_1 and A_2) and two environmental conditions (E_1 and E_2) of an environmental variable E. This setting leads to six possible classes—combinations of genotypes and environments—and thus to six phenotypic expectations. Those values are gathered in the column-vector of genotypic values, $\mathbf{G} = (G_{ijk})$, where the subscripts indicate genotype A_jA_k at environment E_i.

The genotypic values can be expressed in terms of genetic effects by means of regression model:

$$\mathbf{G} = \mathbf{N}_{\mu}\mu + \mathbf{N}_{\varepsilon}\boldsymbol{\varepsilon} + \mathbf{N}_{\alpha}\boldsymbol{\alpha} + \mathbf{N}_{\delta}\boldsymbol{\delta} + \mathbf{N}_{\alpha\varepsilon}\boldsymbol{\alpha\varepsilon} + \boldsymbol{\delta\varepsilon}, \tag{6.10}$$

in which the explanatory variables are the mean phenotype μ; the environmental effect, $\boldsymbol{\varepsilon} = \boldsymbol{\upsilon}_1 = (\varepsilon_1, \varepsilon_2)^{\mathrm{T}}$; the genetic additive effect, $\boldsymbol{\alpha} = \boldsymbol{\upsilon}_2 = (\alpha_1, \alpha_2)^{\mathrm{T}}$; the dominance effect, $\boldsymbol{\delta} = \boldsymbol{\upsilon}_3 = (\delta_{11}, \delta_{12}, \delta_{22})^{\mathrm{T}}$; and the additive-by-environment effect, $\boldsymbol{\alpha\varepsilon} = \boldsymbol{\upsilon}_4 = (\alpha e_{11}, \alpha e_{12}, \alpha e_{21}, \alpha e_{22})^{\mathrm{T}}$, and the residual term is the dominance-by-environment effect, $\boldsymbol{\delta\varepsilon} = \boldsymbol{\eta}_4 = (\delta\varepsilon_{ijk})$. Concerning the design matrices, assuming (as introduced above) that $\mathbf{1}^{(m)}$ is a column vector of m ones, $\mathbf{I}^{(n)}$ is an identity matrix of dimension n, $\mathbf{N}$ is the design matrix of the one-locus additive component (Expression 6.1), and $\otimes$ is the Kronecker product; then, Expression 6.10 gets completed by:

$$\mathbf{N}_{\mu} = \mathbf{N}_0 = \mathbf{1}^{(2)} \otimes \mathbf{1}^{(3)}, \mathbf{N}_{\varepsilon} = \mathbf{N}_1 = \mathbf{I}^{(2)} \otimes \mathbf{1}^{(3)}, \mathbf{N}_{\alpha} = \mathbf{N}_2 = \mathbf{1}^{(2)} \otimes \mathbf{N}, \mathbf{N}_{\delta} = \mathbf{N}_3 = \mathbf{1}^{(2)} \otimes \mathbf{I}^{(3)}, \text{and } \mathbf{N}_{\alpha\varepsilon} = \mathbf{N}_4 = \mathbf{I}^{(2)} \otimes \mathbf{N}, \tag{6.11}$$

where it is key to keep in mind that design matrices $\mathbf{N}_c$, $c = 1,\ldots,4$ are different from the previous case (Expressions 6.5, 6.7–6.9).

Expression 6.11 adheres to the following logic. The right-hand-side of each Kronecker product is related to the genetic contribution, while the left-hand-side is related to the environmental contribution. Whenever one side is not represented in the subscripts, it contributes to the corresponding design matrix through a vector of ones, whose dimension is given by the possible alternatives of that side (two environmental alternatives and three genotypes). Whenever one side is represented in the subscripts, it contributes to the corresponding design matrix through a matrix. For the environmental side and the dominance component of the genetic side, those matrices are identity matrices (of dimension two and three respectively, as justified just above). For the additive component, the matrix is $\mathbf{N}$.

The solution to this new regression framework (Expression 6.10), with these new design matrices (Expression 6.11), has to be obtained in the same way as in the previous case (Expression 6.3), i.e., sequentially. Indeed, the solutions to regression framework in Expression 6.10 can still be obtained using Expression 6.4, provided that we use the design matrices in Expression 6.11 as mentioned right above, and that we now adjust the dimensions of the identity matrices in $\mathbf{M}_c = \mathbf{I}^{(6)} - \mathbf{N}_c\tilde{\mathbf{H}}_c$ (instead of $\mathbf{M}_c = \mathbf{I}^{(9)} - \mathbf{N}_c\tilde{\mathbf{H}}_c$) and $\boldsymbol{\eta}_0 = \mathbf{G} - \mathbf{1}^{(6)}\mu$ (instead of $\boldsymbol{\eta}_0 = \mathbf{G} - \mathbf{1}^{(9)}\mu$). Thus, although biologically different, departures from linkage equilibrium and from random associations between genes and environments are mathematically analogous and, therefore, can be solved using a very similar procedure.

6.2.4 Increasingly Complex Gene-Environment Interplays

Expressions 6.10 and 6.11 can be extended to more complex genetic architectures and environmental conditions in the same way as explained for extending Expression 6.3 to more complex genetic architectures above. For instance, a multiallelic locus can be implemented by just adjusting the dimensions to the corresponding multiallelic one-locus design matrices. For the particular case of three alleles (Expression 5.14), the design matrix of the population mean increases from $\mathbf{1}^{(3)}$ in the biallelic case to $\mathbf{1}^{(6)}$, the dimension of the design matrix of the additive component increases from 3 × 2 in the biallelic case (see Expression 6.1) to 6 × 3, and thus the identity matrix of the dominance component increases from $\mathbf{I}^{(3)}$ in the biallelic case (see text right above Expression 6.7) to $\mathbf{I}^{(6)}$. This way, the design matrices of the gene-environment model with three alleles become:

$$\mathbf{N}_{\mu}=\mathbf{N}_0=\mathbf{1}^{(2)}\otimes\mathbf{1}^{(6)},\mathbf{N}_{\varepsilon}=\mathbf{N}_1=\mathbf{I}^{(2)}\otimes\mathbf{1}^{(6)},\mathbf{N}_{\alpha}=\mathbf{N}_2=\mathbf{1}^{(2)}\otimes\mathbf{N},\mathbf{N}_{\delta}$$
$$=\mathbf{N}_3=\mathbf{1}^{(2)}\otimes\mathbf{I}^{(6)},\text{ and }\mathbf{N}_{\alpha\varepsilon}=\mathbf{N}_4=\mathbf{I}^{(2)}\otimes\mathbf{N}, \tag{6.12}$$

where it is understood that **N** stands in this case for the aforementioned three-allele design matrix of the additive component.

The extension to more conditions of the environmental variable can also be performed straightforwardly. For instance, increasing the number of possible conditions from two to three (e.g., individuals that can develop under three different average temperatures) leads to the following design matrices:

$$\mathbf{N}_{\mu}=\mathbf{N}_0=\mathbf{1}^{(3)}\otimes\mathbf{1}^{(3)},\mathbf{N}_{\varepsilon}=\mathbf{N}_1=\mathbf{I}^{(3)}\otimes\mathbf{1}^{(3)},\mathbf{N}_{\alpha}=\mathbf{N}_2=\mathbf{1}^{(3)}\otimes\mathbf{N},\mathbf{N}_{\delta}$$
$$=\mathbf{N}_3=\mathbf{1}^{(3)}\otimes\mathbf{I}^{(3)},\text{ and }\mathbf{N}_{\alpha\varepsilon}=\mathbf{N}_4=\mathbf{I}^{(3)}\otimes\mathbf{N}, \tag{6.13}$$

where this time we have just exchanged $\mathbf{1}^{(2)}$ and $\mathbf{I}^{(2)}$ by $\mathbf{1}^{(3)}$ and $\mathbf{I}^{(3)}$, respectively. Note that the changes from Expression 6.11 to Expressions 6.12 and 6.13 can easily be performed simultaneously, thus leading to a model with three alleles and three environmental conditions. More in general, following the same logic, we can extend the model to k_1 alleles and k_2 environmental conditions.

Regression model in Expression 6.10 can also be extended to more loci and/or more environmental variables. For instance, the extension to two biallelic loci (with arbitrary epistasis and arbitrary departures from linkage equilibrium), i.e., to completely arbitrary gene-gene-environment interplay, can be done using Expression 6.3 and adding up the emerging gene-interaction terms. The resulting regression framework is:

$$\mathbf{G}=\mathbf{N}_{\mu}\,\mu+\mathbf{N}_{\varepsilon}\,\boldsymbol{\varepsilon}+\mathbf{N}_{\alpha}\,\boldsymbol{\alpha}+\mathbf{N}_{\delta}\,\boldsymbol{\delta}+\mathbf{N}_{\alpha\alpha}\boldsymbol{\alpha\alpha}+(\mathbf{N}_{\delta\alpha}|\mathbf{N}_{\alpha\delta})(\boldsymbol{\delta\alpha}|\boldsymbol{\alpha\delta})^{\mathrm{T}}+\mathbf{N}_{\delta\delta}\boldsymbol{\delta\delta}$$
$$+\mathbf{N}_{\alpha\varepsilon}\boldsymbol{\alpha\varepsilon}+\mathbf{N}_{\delta\varepsilon}\boldsymbol{\delta\varepsilon}+\mathbf{N}_{\alpha\alpha\varepsilon}\,\boldsymbol{\alpha\alpha\varepsilon},+(\mathbf{N}_{\delta\alpha\varepsilon}|\mathbf{N}_{\alpha\delta\varepsilon})(\boldsymbol{\delta\alpha\varepsilon}|\boldsymbol{\alpha\delta\varepsilon})^{\mathrm{T}}+\boldsymbol{\delta\delta\varepsilon}. \tag{6.14}$$

In general, all design matrices in this expression can be derived along the lines of the ones of Expression 6.10—i.e., as in Expression 6.11. In particular:

$$
\begin{aligned}
\mathbf{N}_{\mu} &= \mathbf{N}_0 = \mathbf{1}^{(2)} \otimes \mathbf{1}^{(9)}, \mathbf{N}_{\varepsilon} = \mathbf{N}_1 = \mathbf{I}^{(2)} \otimes \mathbf{1}^{(9)}, \mathbf{N}_{\alpha} = \mathbf{N}_2 = \mathbf{1}^{(2)} \otimes \mathbf{N}_{\alpha}^{2l}, \mathbf{N}_{\delta} \\
&= \mathbf{N}_3 = \mathbf{1}^{(2)} \otimes \mathbf{N}_{\delta}^{2l}, \mathbf{N}_{\alpha\alpha} = \mathbf{N}_4 = \mathbf{1}^{(2)} \otimes \mathbf{N}_{\alpha\alpha}^{2l}, (\mathbf{N}_{\delta\alpha}|\mathbf{N}_{\alpha\delta}) = \mathbf{N}_5 \\
&= (\mathbf{1}^{(2)} \otimes \mathbf{N}_{\delta\alpha}^{2l}|\mathbf{1}^{(2)} \otimes \mathbf{N}_{\alpha\delta}^{2l}), \mathbf{N}_{\delta\delta} = \mathbf{N}_6 = \mathbf{1}^{(2)} \otimes \mathbf{I}^{(9)}, \mathbf{N}_{\alpha\varepsilon} = \mathbf{N}_7 \\
&= \mathbf{I}^{(2)} \otimes \mathbf{N}_{\alpha}, \mathbf{N}_{\delta\varepsilon} = \mathbf{N}_8 = \mathbf{I}^{(2)} \otimes \mathbf{N}_{\delta}, \mathbf{N}_{\alpha\alpha\varepsilon} = \mathbf{N}_9 \\
&= \mathbf{I}^{(2)} \otimes \mathbf{N}_{\alpha\alpha}, \text{and } (\mathbf{N}_{\delta\alpha\varepsilon}|\mathbf{N}_{\alpha\delta\varepsilon}) = \mathbf{N}_{10} = (\mathbf{I}^{(2)} \otimes \mathbf{N}_{\delta\alpha}|\mathbf{I}^{(2)} \otimes \mathbf{N}_{\alpha\delta}),
\end{aligned}
\tag{6.15}
$$

where matrices $\mathbf{N}_{\alpha}^{2l}$, $\mathbf{N}_{\delta}^{2l}$, $\mathbf{N}_{\alpha\alpha}^{2l}$, $\mathbf{N}_{\delta\alpha}^{2l}$, and $\mathbf{N}_{\alpha\delta}^{2l}$ can be taken from Expressions 6.5, 6.7, 6.8, and 6.9, by just taking into account that they lack the superscript (indicating "two loci") in those expressions. Although Expression 6.15 may look difficult to read at first sight, it follows the same logic as Expression 6.11. Since Expression 6.15 has more components than Expression 6.11, there are also more design matrices of previous models to take into account for building it. In particular, only the design matrix of the additive component of the one-locus model Expression 6.3, **N**, was necessary to build Expression 6.11, whereas for Expression 6.15 we need all design matrices of the two-locus model (Expressions 6.3, 6.4–6.9), $\mathbf{N}_{\alpha}^{2l}$, $\mathbf{N}_{\delta}^{2l}$, $\mathbf{N}_{\alpha\alpha}^{2l}$, $\mathbf{N}_{\delta\alpha}^{2l}$, and $\mathbf{N}_{\alpha\delta}^{2l}$, as mentioned just above.

The solution to Expression 6.14 with the design matrices in Expression 6.14 can be obtained in the same way as for the previous cases. In other words, the solution can be obtained through Expression 6.4, although in the present case there are ten (instead of four) sequential steps to perform, i.e., $c = 1, \ldots, 10$. As in some of the previous cases, Expression 6.4 has to be modified by Expression 6.6 whenever necessary.

To end up with, also the number of environmental variables considered may increase (to, for instance, temperature and humidity, or to whether a person is a smoker or not and has a sedentary lifestyle or not), while still attaining generality in what regards the whole gene-environment interplay. In the case of two environmental factors (i.e., gene-environment-environment interaction) that can display two environmental conditions each, the regression model becomes:

$$
\begin{aligned}
\mathbf{G} = \mathbf{N}_{\mu}\mu + \mathbf{N}_{\varepsilon 1}\boldsymbol{\varepsilon}_1 + \mathbf{N}_{\varepsilon 2}\boldsymbol{\varepsilon}_2 + \mathbf{N}_{\alpha}\boldsymbol{\alpha} + \mathbf{N}_{\delta}\boldsymbol{\delta} + \mathbf{N}_{\alpha\varepsilon 1}\boldsymbol{\alpha\varepsilon}_1 + \mathbf{N}_{\alpha\varepsilon 2}\boldsymbol{\alpha\varepsilon}_2 \\
+ \mathbf{N}_{\delta\varepsilon 1}\boldsymbol{\delta\varepsilon}_1 + \mathbf{N}_{\delta\varepsilon 2}\boldsymbol{\delta\varepsilon}_2 + \mathbf{N}_{\varepsilon\varepsilon}\boldsymbol{\varepsilon\varepsilon} + \mathbf{N}_{\alpha\varepsilon\varepsilon}\boldsymbol{\alpha\varepsilon\varepsilon} + \boldsymbol{\delta\varepsilon\varepsilon}.
\end{aligned}
\tag{6.16}
$$

The design matrices that complete Expression 6.16 can be built with the same logic as for the previous cases (Expressions 6.13 and 6.15), just taking into account that an extra environmental factor implies an extra factor in the Kronecker, products. Thus the design matrices in Expression 6.16 become:

$$\mathbf{N}_{\mu} = \mathbf{N}_0 = \mathbf{1}^{(2)} \bigotimes \mathbf{1}^{(2)} \bigotimes \mathbf{1}^{(3)}, \mathbf{N}_{\varepsilon 1} = \mathbf{N}_1 = \mathbf{I}^{(2)} \bigotimes \mathbf{1}^{(2)} \bigotimes \mathbf{1}^{(3)}, \mathbf{N}_{\varepsilon 2}$$
$$= \mathbf{N}_2 = \mathbf{1}^{(2)} \bigotimes \mathbf{I}^{(2)} \bigotimes \mathbf{1}^{(3)}, \mathbf{N}_{\alpha} = \mathbf{N}_3 = \mathbf{1}^{(2)} \bigotimes \mathbf{1}^{(2)} \bigotimes \mathbf{N}, \mathbf{N}_{\delta} = \mathbf{N}_4$$
$$= \mathbf{1}^{(2)} \bigotimes \mathbf{1}^{(2)} \bigotimes \mathbf{I}^{(3)}, \mathbf{N}_{\alpha\varepsilon 1} = \mathbf{N}_5 = \mathbf{I}^{(2)} \bigotimes \mathbf{1}^{(2)} \bigotimes \mathbf{N}, \mathbf{N}_{\alpha\varepsilon 2} = \mathbf{N}_6$$
$$= \mathbf{1}^{(2)} \bigotimes \mathbf{I}^{(2)} \bigotimes \mathbf{N}, \mathbf{N}_{\delta\varepsilon 1} = \mathbf{N}_7 = \mathbf{I}^{(2)} \bigotimes \mathbf{1}^{(2)} \bigotimes \mathbf{I}^{(3)}, \mathbf{N}_{\delta\varepsilon 2} = \mathbf{N}_8$$
$$= \mathbf{1}^{(2)} \bigotimes \mathbf{I}^{(2)} \bigotimes \mathbf{I}^{(3)}, \mathbf{N}_{\varepsilon\varepsilon} = \mathbf{N}_9 = \mathbf{I}^{(2)} \bigotimes \mathbf{I}^{(2)} \bigotimes \mathbf{1}^{(3)}, \mathbf{N}_{\alpha\varepsilon\varepsilon} = \mathbf{N}_{10}$$
$$= \mathbf{I}^{(2)} \bigotimes \mathbf{I}^{(2)} \bigotimes \mathbf{N}, \qquad (6.17)$$

where **N** is the design matrix of the one-locus additive component (Expression 6.1).

The solution to Expressions 6.16 and 6.17 can be obtained just as for the case of Expressions 6.14 and 6.15, i.e., using Expression 6.4 with $c = 1, \ldots, 10$, and applying the alternative in Expression 6.6 whenever necessary. By applying the logic of the previous extensions, several of them can be made simultaneously to model, say, a gene-gene-environment-environment interaction with a biallelic locus and a three-allele locus and two environments that can display four and two alternatives, respectively.

6.3 ARNOIA: NOIA Beyond Random Associations

The COIA regression framework (Expressions 6.1–6.17) provides a theory of genetic and environmental parameters accounting for arbitrary associations between/among factors (whether loci or environments or both) at the population level, thus providing a GP map beyond correlations (i.e., beyond linkage equilibrium and non-random associations between genes and environments). In other words, COIA (performed in addition to the theory presented in the previous chapter) provides a GP appropriate to perform an orthogonal decomposition of the genotypic values with general population frequencies, thus accounting in particular for correlations between factors. It is indeed for this reason that those expressions are labeled as the correlation-wise orthogonal interactions (COIA) regression framework (Álvarez-Castro 2020).

As the analysis at the individual level does not have to see with population frequencies, we had already obtained a general formulation of NOIA at the individual level in Chap. 4. Then, in Chap. 5, we obtained a formulation of NOIA at the population level that was general under linkage equilibrium and non-random associations between genes and environments. Thus, in order to also attain a general formulation of NOIA at the population level here, we just need to use COIA to extend NOIA to non-random associations between/amongst factors (genes and environments). Such extension becomes an association-resolved NOIA, which is thus referred to as ARNOIA (Álvarez-Castro 2020).

The implementation of COIA within the NOIA framework implies to develop genetic-effects design matrices (to be precise, genetic-and-environmental-effects design matrices, whenever gene-environment interactions come into play), **S**, providing an orthogonal decomposition of the genotypic values as **G** = **SE**, based on the parameters obtained with COIA. Such decomposition based on **S** is obtained via its inverse, $\mathbf{S}^{-1}$. Since $\mathbf{E} = \mathbf{S}^{-1}\mathbf{G}$, the scalars of this latest matrix are the scalars of the genetic effects (in vector $\mathbf{E} = (\mu, \alpha, \delta, \ldots)^T$) expressed as linear combinations of the genotypic values (in vector **G**), and they can be obtained as follows.

First, the scalars of the mean phenotype are just the genotypic frequencies. Then, the genetic effects themselves can be built up based on the parameters obtained with COIA (Expressions 6.1–6.17). Those parameters have to be combined in the particular ways leading to the genetic effects at the population level to be well defined—so that the correct decomposition of the genotypic values and of the genetic variance can then be obtained from them. The ways in which those parameters have to be combined can be deduced using the corresponding $\mathbf{S}^{-1}$ matrix at the individual level—dealt with in Chap. 4—as a template. In particular, the non-nil scalars of the matrix at the individual level indicate which scalars have to be used to combine the parameters obtained with COIA for obtaining the genetic effects and, therefore, to obtain the rows of the $\mathbf{S}^{-1}$ matrix at the population level. To illustrate this, the cases of two biallelic autosomal loci and interaction between one biallelic locus and two environmental conditions are developed in detail below.

6.3.1 Two Biallelic Autosomal Loci

Let us first have a look at the single locus $\mathbf{S}^{-1}$ matrix at the individual level (Expression 4.4):

$$\begin{pmatrix} 1 & 0 & 0 \\ -\frac{1}{2} & 0 & \frac{1}{2} \\ -\frac{1}{2} & 1 & -\frac{1}{2} \end{pmatrix}. \tag{6.18}$$

Note that this matrix provides a clue of how the genetic effects of one biallelic locus at the population level are defined. First, the additive genetic effect is $\alpha = \alpha_2 - \alpha_1$ (see, e.g., Expression 2.9). The coefficients of α_1 and α_2 (minus one and one, respectively) are the non-nil scalars of the second row of the matrix in Expression 6.18, multiplied by two. Then, the dominance genetic effect is defined at the individual level through the third row of Expression 6.18 in the very same way as at the population level (see Expression 5.11), i.e., $\delta = \delta_{12}–(\delta_{11} + \delta_{22})/2$. The third row of the matrix in Expression 6.18 provides the scalars of δ_{11}, δ_{12}, and δ_{12} ($-\frac{1}{2}$, 1, and $-\frac{1}{2}$). Overall, matrix (6.18) provides the scalars to obtain the genetic effects from the corresponding estimates obtained using the regression framework (Expression 6.1), by just multiplying the scalars of the additive effect times two.

Then, the same logic enables us to also build the $\mathbf{S}^{-1}$ matrix of more complex models at the population level. With two biallelic loci, the $\mathbf{S}^{-1}$ square matrix is of dimension nine, the column vector of genetic effects being $\mathbf{E} = (\mu, \alpha^1, \delta^1, \alpha^2, \alpha\alpha, \delta\alpha,$

δ^2, $\alpha\delta$, $\delta\delta)^{\mathrm{T}}$. There is a row of $\mathbf{S}^{-1}$ for each scalar of vector **E**. These rows must be filled with the coefficients of that scalar expressed as a linear combination of the genotypic values, G_{ijkl}. The aim is thus to define each of the scalars of **E** as linear combinations of the genotypic values. The first row, corresponding to scalar μ (the mean phenotypic value), must thus be filled with the genotypic frequencies, p_{ijkl}, since $\mu = \Sigma p_{ijkl}G_{ijkl}$.

Next, we must obtain all remaining scalars of **E** in terms of linear combinations of the genotypic values. To do so, we shall first obtain the $\mathbf{S}^{-1}$ matrix at the individual level, to be used (as explained above) as a template to get to the desired definitions of the genotypic values. The single locus $\mathbf{S}^{-1}$ matrix at the individual level is that of Expression 6.18. Then, following the indications under the "Multiple loci" heading of Chap. 4, the two-locus $\mathbf{S}^{-1}$ matrix at the individual level is $\mathbf{S}^{-1} \otimes \mathbf{S}^{-1}$ (which, incidentally, could also be computed as $(\mathbf{S} \otimes \mathbf{S})^{-1}$), expanding to:

$$\begin{pmatrix} 1 & 0 & 0 & 0 & 0 & 0 & 0 & 0 & 0 \\ -\frac{1}{2} & 0 & \frac{1}{2} & 0 & 0 & 0 & 0 & 0 & 0 \\ -\frac{1}{2} & 1 & -\frac{1}{2} & 0 & 0 & 0 & 0 & 0 & 0 \\ -\frac{1}{2} & 0 & 0 & 0 & 0 & 0 & \frac{1}{2} & 0 & 0 \\ \frac{1}{4} & 0 & -\frac{1}{4} & 0 & 0 & 0 & -\frac{1}{4} & 0 & \frac{1}{4} \\ \frac{1}{4} & -\frac{1}{2} & \frac{1}{4} & 0 & 0 & 0 & -\frac{1}{4} & \frac{1}{2} & -\frac{1}{4} \\ -\frac{1}{2} & 0 & 0 & 1 & 0 & 0 & -\frac{1}{2} & 0 & 0 \\ \frac{1}{4} & 0 & -\frac{1}{4} & -\frac{1}{2} & 0 & \frac{1}{2} & \frac{1}{4} & 0 & -\frac{1}{4} \\ \frac{1}{4} & -\frac{1}{2} & \frac{1}{4} & -\frac{1}{2} & 1 & -\frac{1}{2} & \frac{1}{4} & -\frac{1}{2} & \frac{1}{4} \end{pmatrix}. \tag{6.19}$$

Now, the marginal effects in vector $\mathbf{E} = (\mu, \alpha^1, \delta^1, \alpha^2, \alpha\alpha, \delta\alpha, \delta^2, \alpha\delta, \delta\delta)^{\mathrm{T}}$ are in positions 2, 3, 4 and 7, thus determined by the corresponding rows of the matrix in Expression 6.19 as:

$$\alpha^l = \left(\alpha_2^l - \alpha_1^l\right), \delta^l = \delta_{12}^l - \left(\delta_{11}^l + \delta_{22}^l\right)/2, l = 1, 2. \tag{6.20}$$

The loci are indicated as superscripts, since we have used subscripts above to indicate the different parameters to estimate in the regressions in Expressions 6.1, 6.5, 6.7–6.9. In previous chapters, we have used superscripts as indicators for pairs of alleles. If loci and pairs of alleles have to be indicated at the same time, then compound superscripts can be used. For instance, $\alpha^{2.13}$ would be the additive effect of allele substitutions between alleles "1" and "3" of locus "2"—i.e., alleles A_1^2 and A_3^2.

In any case, Expression 6.20 comes as no surprise as we have been defining the additive effects in this way throughout Chaps. 2 and 5, and we have also defined the dominance effects in this way previously, particularly in the X-linked and the imprinting models in Chap. 5. Indeed, we would not even need the matrix at the individual level for getting to these definitions. Nevertheless, the definitions of the marginal genetic effects are made in terms of the genetic parameters (Expression 6.20), which have been obtained above as linear combinations of the genotypic values. Thus, they provide also linear combinations of the genotypic values, whose coefficients provide in their turn the corresponding rows of the $\mathbf{S}^{-1}$ matrix at the population level.

Next, we jump all the way to the highest-level interaction term, $\delta\delta$, in the last position of vector **E**—position 9. In this case, the last row of the matrix at the individual level (Expression 6.19) directly provides the coefficients of $\delta\delta$ as a linear combination of the genotypic values. This is because the genetic effect of the highest order interaction is the same at the individual level and at the population level. Thus, this row must be brought as it is from the matrix at the individual level (Expression 6.19) to the matrix at the population level. This is also as already applied throughout Chap. 5—although in this case, as opposed to the marginal effects, the matrix at the individual level (Expression 6.19) is needed.

To end up with, we proceed with the remaining interaction effects (all interaction effects implying more than one locus, except from the highest-level interaction, $\delta\delta$), i.e., $\alpha\alpha$, $\delta\alpha$, and $\alpha\delta$. Since they are in positions 5, 6, and 8 of column vector **E**, respectively, they are defined from the corresponding parameters obtained using COIA (in particular, Expressions 6.8 and 6.9), using rows 5, 6, and 8 of the matrix in Expression 6.19 as template. In particular, the non-nil scalars of those rows must be used as coefficients after multiplying each row times two as many times as additive effects are involved (two additive effects in the fifth row, of $\alpha\alpha$, and one additive effect in the sixth and the eight rows, of $\delta\alpha$, and $\alpha\delta$, respectively). Thus, definitions of $\alpha\alpha$, $\delta\alpha$, and $\alpha\delta$ are, in particular:

$$\begin{aligned}&\alpha\alpha = \alpha\alpha_{11} - \alpha\alpha_{12} - \alpha\alpha_{21} + \alpha\alpha_{22}, \delta\alpha = ½\, \delta\alpha_{111} - \delta\alpha_{121} + ½\delta\alpha_{221}\\&\quad - ½\, \delta\alpha_{112} + \delta\alpha_{122} - ½\, \delta\alpha_{222}, \text{and } \alpha\delta = ½\, \alpha\delta_{111} - ½\, \alpha\delta_{211} - \alpha\delta_{112}\\&\quad + \alpha\delta_{212} + ½\, \alpha\delta_{122} - ½\, \alpha\delta_{222}.\end{aligned} \tag{6.21}$$

Then again, since the parameters used to define $\alpha\alpha$, $\delta\alpha$, and $\alpha\delta$ in this expression have been obtained above as linear combinations of the genotypic values, the genetic effects $\alpha\alpha$, $\delta\alpha$, and $\alpha\delta$ are themselves also provided here (Expression 6.21) as linear combinations of the genotypic values, and the coefficients of those linear combinations are the scalars of the corresponding rows of the $\mathbf{S}^{-1}$ matrix at the population level.

6.3.2 One Biallelic Locus and Two Environmental Conditions

In the case of gene-environment interaction, we need two different $\mathbf{S}^{-1}$ square matrices at the individual level—the inverse of the genetic-effects design matrix and the inverse of the environmental-effects design matrix. The first one has been given in Expression 6.19, and the second one (provided in Expression 4.18) is:

$$\begin{pmatrix} 1 & 0 \\ -1 & 1 \end{pmatrix}. \tag{6.22}$$

The $\mathbf{S}^{-1}$ matrix at the individual level needed as a template for obtaining the genetic/environmental effects (and, thus, the $\mathbf{S}^{-1}$ matrix) at the population level can thus for this case be computed as the Kronecker product of the matrices in Expressions 6.19 and 6.22, which is:

$$\begin{pmatrix} 1 & 0 & 0 & 0 & 0 & 0 \\ -\tfrac{1}{2} & 0 & \tfrac{1}{2} & 0 & 0 & 0 \\ -\tfrac{1}{2} & 1 & -\tfrac{1}{2} & 0 & 0 & 0 \\ -1 & 0 & 0 & 1 & 0 & 0 \\ \tfrac{1}{2} & 0 & -\tfrac{1}{2} & -\tfrac{1}{2} & 0 & \tfrac{1}{2} \\ \tfrac{1}{2} & -1 & \tfrac{1}{2} & -\tfrac{1}{2} & 1 & -\tfrac{1}{2} \end{pmatrix}. \tag{6.23}$$

The vector of genetic/environmental effects is $\mathbf{E} = (\mu, \alpha, \delta, \varepsilon, \alpha\varepsilon, \delta\varepsilon)^{\mathrm{T}}$. The first row of the inverse of the genetic/environmental-effects design matrix—the one of the population mean phenotype—is given, as in the previous case, by the genotypic frequencies. The marginal effects are in this case, α, δ, and ε, in positions 2 to 4, respectively. The additive and dominance effects, α and δ, are defined by their corresponding rows in the matrix in Expression 6.23 just as in Expression 6.20, with no need for superscripts in the present case. As for the environmental effect, ε, can be obtained using the fourth row of that matrix just as:

$$\varepsilon = \varepsilon_2 - \varepsilon_1. \tag{6.24}$$

Thus, as in the case of the two autosomal biallelic loci right above, all marginal effects in this case are defined as in previous chapters (for ε, see the "Gene-environment interaction" section in Chap. 5).

The last row in the matrix in Expression 6.23 directly provides, as in the previous case, the last row of the $\mathbf{S}^{-1}$ matrix at the population level—the one for $\delta\varepsilon$ in the present case. With this, one only genetic effect remains to be defined in this case, $\alpha\varepsilon$. Since this genetic effect is at the penultimate position of the vector of genetic effects, $\mathbf{E}$, it can be obtained using the penultimate row of the $\mathbf{S}^{-1}$ matrix at the individual level (Expression 6.23) times two (since one additive effect is involved), as:

$$\alpha\varepsilon = \alpha\varepsilon_{11} - \alpha\varepsilon_{12} - \alpha\varepsilon_{21} + \alpha\varepsilon_{22}. \tag{6.25}$$

The genetic parameters required in the previous definitions can be obtained using COIA, particularly with Expressions 6.10 and 6.11. Then, also as in the previous case, the definitions enable us to express the genetic effects as combinations of the genotypic values, whose coefficients provide, in their turn, the corresponding rows of the desired $\mathbf{S}^{-1}$ matrix at the population level.

6.4 Key Milestone Reached

Fisher (1918) considered epistasis in his seminal work on models of genetic effects but did not find it consequential for the outcome of selection of quantitative traits. As briefly discussed in Chap. 3, Fisher's view became controversial, with Wright taking the lead on the opposing side (Provine 1986). Epistasis was eventually implemented within Fisher's (1918) theoretical framework (Cockerham 1954; Kempthorne 1954). As also commented earlier (in Chap. 5), the practical applicability of that accomplishment was severely limited by the data and the computational power available at that moment, when the exponential growth of molecular genetics had barely been triggered.

Fisher and Wright's controversy on epistasis resurfaced with renewed strength when gene-mapping experiments became feasible—after Lander and Botstein's (1989) kick-off of quantitative trait loci (QTL) analysis. Amongst the major concerns from that point on were adapting the mapping procedures to detect epistasis and developing procedures to analyze its roles in selection and evolution (see e.g. Wolf et al. 2000; Hansen 2013; Moore and Williams 2015). Although deciphering epistasis proved cunning, the view that gene-interactions in general can hardly be avoided in the study of selection and evolution has become reinforced.

One approach to detect epistasis exploits the linkage disequilibrium footprint of selection on epistatic genetic architectures (Corbett-Detig et al. 2013). In contrast, linkage disequilibrium hinders the analyses of epistatic genetic architectures using conventional methods—not implementing departures from linkage equilibrium properly. Such drawback was approached under different perspectives (e.g., Yang 2004; Mao et al. 2006; Wang and Zeng 2006, 2009; Hill and Mäki-Tanila 2015). In general, confounding effects of the two phenomena have been commented in different ways at different times (see e.g. Cockerham 1954; Zan et al. 2018). However, attaining a theory of genetic effects properly accounting for arbitrary departures from linkage equilibrium was deemed discouraging (e.g., Wang and Zeng 2006; Hill and Mäki-Tanila 2015; Vitezica et al. 2017). Overall, it could be said that epistasis has not only given rise to heated controversy, but probably it has even become surrounded by a certain aura of mystery.

Interestingly, the apparent muddle about formulating a theory of genetic effects satisfactorily combining epistasis (i.e., gene-gene interaction) and departures from linkage equilibrium (i.e., correlation between/among allele frequencies of different loci) contrasts with the clear-cut way in which the analogous challenge of combining gene-environment interaction and gene-environment correlation has been faced. In particular, in relation with mental health, it has been acutely urged that "[i]dentifying which form of gene-environment interplay contributes to a particular disorder or behavior is absolutely crucial in order to select suitable intervention efforts" and that theory enabling to properly analyze the two phenomena jointly is required for "ensuring that the outcomes of one do not bias the effects of the other" (Assary et al. 2020).

As explained above, the mathematical procedure to develop a GP at the population level properly implementing the gene-gene and the gene-environment interplays is basically the same—COIA (Álvarez-Castro 2020). In brief, it consists in applying the Kronecker product at the level of the design matrices of the regression framework to obtain the genetic/environmental parameters, instead of applying it at the level of the genetic-effects design matrices. Thus, COIA addresses multiple factors (loci and environments) in an integrated manner from the outset (Álvarez-Castro 2020), rather than initially considering them one by one, as previous proposals with matrix notation did—particularly, the G2A model (Zeng et al. 2005) and the original proposal of NOIA (Álvarez-Castro and Carlborg 2007).

Then, from the genetic/environmental parameters obtained with COIA, a multiple-factor (loci and environments) genetic-effects design matrix at the population level can be mounted in one stroke. To this aim, interaction genetic effects that became defined automatically in the original NOIA model (through the Kronecker product of genetic-effects design matrices or of their inverses) have to be defined explicitly (using the model at the individual level as a template). Building such genetic-effects design matrix enables the implementation of COIA into the unifying framework also entailing genetic effects at the individual level, i.e., into the NOIA framework. (Álvarez-Castro 2020).

Overall, we have in this chapter upgraded NOIA to properly account for departures from linkage equilibrium and from random associations between genes and environments. This achievement has required a significant methodological addition. The resulting theory can therefore be considered a 2.0 version of NOIA and as such it is convenient to assign a label to it—ARNOIA. In attention to both the methodological developments required and the practical limitations overcome, ARNOIA stands as a relevant milestone, crucial in the achievement of a general GP map.

References

Álvarez-Castro JM (2012) Current applications of models of genetic effects with interactions across the genome. Curr Genomics 13:163–175

Álvarez-Castro JM (2014) Dissecting genetic effects with imprinting. Front Ecol Evol 2:51

Álvarez-Castro JM (2020) Gene-environment interaction in the era of precision medicine—filling the potholes rather than starting to build a new road. Front Genet 11:921

Álvarez-Castro JM, Carlborg Ö (2007) A unified model for functional and statistical epistasis and its application in quantitative trait loci analysis. Genetics 176:1151–1167

Álvarez-Castro JM, Crujeiras RM (2019) Orthogonal decomposition of the genetic variance for epistatic traits under linkage disequilibrium-applications to the analysis of Bateson-Dobzhansky-Muller incompatibilities and sign epistasis. Front Genet 10:54

Álvarez-Castro JM, Yang R-C (2011) Multiallelic models of genetic effects and variance decomposition in non-equilibrium populations. Genetica 139:1119–1134

Assary E, Vincent J, Machlitt-Northen S, Keers R, Pluess M (2020) The role of gene-environment interaction in mental health and susceptibility to the development of plychiatric disorders. In: Teperino R (ed) Beyond our genes. Pathophysiology of gene and environment interaction and epigenetic inheritance. Springer, Cham, pp 117–138

Cheverud JM, Routman EJ (1995) Epistasis and its contribution to genetic variance components. Genetics 139:1455–1461

Cockerham CC (1954) An extension of the concept of partitioning hereditary variance for analysis of covariances among relatives when epistasis is present. Genetics 39:859–882
Corbett-Detig RB, Zhou J, Clark AG, Hartl DL, Ayroles JF (2013) Genetic incompatibilities are widespread within species. Nature 504:135–137
Fisher RA (1918) The correlation between relatives on the supposition of Mendelian inheritance. Trans Roy Soc Edinburgh 52:339–433
Giesel JT (1977) A model of functional epistasis and linkage disequilibrium in populations with overlapping generations. Genetics 86:679–686
Hansen TF (2013) Why epistasis is important for selection and adaptation. Evolution 67:3501–3511
Hansen TF, Wagner GP (2001) Modeling genetic architecture: a multilinear theory of gene interaction. Theor Popul Biol 59:61–86
Hill WG, Mäki-Tanila A (2015) Expected influence of linkage disequilibrium on genetic variance caused by dominance and epistasis on quantitative traits. J Anim Breed Genet 132:176–186
Jana S (1972) Simulation of quantitative characters from qualitatively acting genes II. Orthogonal subdivision of hereditary variance in two-locus genetic systems. Theor Appl Genet 42:119–124
Kempthorne O (1954) The correlation between relatives in a random mating population. Proc R Soc Lond B Biol Sci 143:102–113
Lander ES, Botstein D (1989) Mapping mendelian factors underlying quantitative traits using RFLP linkage maps. Genetics 121:185–199
Lewontin RC, Kojima K (1960) The evolutionary dynamics of complex polymorphisms. Evolution 14:458–472
Ma J, Xiao F, Xiong M, Andrew AS, Brenner H, Duell EJ, Haugen A, Hoggart C, Hung RJ, Lazarus P, Liu C, Matsuo K, Mayordomo JI, Schwartz AG, Staratschek-Jox A, Wichmann E, Yang P, Amos CI (2012) Natural and orthogonal interaction framework for modeling gene-environment interactions with application to lung cancer. Hum Hered 73:185–194
Mao Y, London NR, Ma L, Dvorkin D, Da Y (2006) Detection of SNP epistasis effects of quantitative traits using an extended Kempthorne model. Physiol Genomics 28:46–52
Moore JH (2005) A global view of epistasis. Nat Genet 37:13–14
Moore JH, Williams SM (2005) Traversing the conceptual difice between biological and statistical epistasis: systems biology and a more modern synthesis. ByoEssays 27:637–646
Moore JH, Williams SM (eds) (2015) Epistasis: methods and protocols. Springer, New York
Phillips PC (2008) Epistasis—the essential role of gene interactions in the structure and evolution of genetic systems. Nat Rev Genet 9:855–867
Provine WB (1986) Sewall wright and evolutionary biology. University of Chicago Press, Chicago
R_Core_Team (2017) R: a language and environment for statistical computing. R Foundation for Statistical Computing, Viena
Seyffert W (1966) Die simulation quantitativer merkmale durch gene mit biochemisch Jefinierbarer Wirkung. Züchter 36:159–162
Slatkin M (2008) Linkage disequilibrium—understanding the evolutionary past and mapping the medical future. Nat Rev Genet 9:477–485
Tiwari HK, Elston RC (1997) Deriving components of genetic variance for multilocus models. Genet Epidemiol 14:1131–1136
Vitezica ZG, Legarra A, Toro MA, Varona L (2017) Orthogonal estimates of variances for additive, dominance, and epistatic effects in populations. Genetics 206:1297–1307
Wang T, Zeng ZB (2006) Models and partition of variance for quantitative trait loci with epistasis and linkage disequilibrium. BMC Genet 7:9
Wang T, Zeng ZB (2009) Contribution of genetic effects to genetic variance components with epistasis and linkage disequilibrium. BMC Genet 10:52
Wolf JB, Brodie ED, Wade MJ (eds) (2000) Epistasis and the evolutionary process. Oxford University Press, New York
Yang R-C (2004) Epistasis of quantitative trait loci under different gene action models. Genetics 167:1493–1505

Yang R-C, Álvarez-Castro JM (2008) Functional and statistical genetic effects with miltiple alleles. Curr Topics Genet 3:49–62
Zan Y, Forsberg SKG, Carlborg O (2018) On the relationship between high-order linkage disequilibrium and epistasis. G3 8:2817–2824
Zeng ZB, Wang T, Zou W (2005) Modeling quantitative trait loci and interpretation of models. Genetics 169:1711–1725

7 Variance Decomposition, Gene Mapping and Average Excesses: Orthogonality in the Spotlight

Abstract

Models of genetic effects—i.e. genotype-to-phenotype (GP) maps—are expressions providing partitions of the genotypic values into components with biological meaning. In the previous chapters, we have achieved an advanced model of genetic effects both at the individual and at the population levels, applicable to very complex genetic architectures and to non-equilibrium populations. Such theory is at the core of quantitative genetics and of a major part of evolutionary biology. This chapter goes beyond that core theory in several ways. First, from the decomposition of the genotypic values, the present chapter moves on to the analysis of the decomposition of the genetic variance. Second, the advantages of using advanced models of genetic effects in mapping experiments are analysed. In particular, a method enabling to obtain adequate (orthogonal) estimates of genetic effects from real data in the face of missing information is here presented. Finally, a more general setting that, beyond the genetic effects, entails the average excesses as well is developed. Overall, crucial developments at the verge of the GP map are here present and discussed. We take advantage of considering these cases together to further clarify the role of orthogonality in models of genetic effects.

7.1 Introduction

As addressed from different perspectives throughout the previous chapters, Fisher (1918) mounted a theory of genetic effects that established the guidelines of quantitative genetics. With that theory, Fisher dissected the genetic basis of a quantitative trait by showing that the resemblance between relatives could be explained in terms of the effects of the hereditary factors they share. In particular, he identified the additive effects as responsible for the resemblance. Fisher's theory thus relies on a partition of the phenotypic effect of each genetic composition (the genotypic values)

J. M. Álvarez-Castro, *Genes, Environments and Interactions*,
https://doi.org/10.1007/978-3-031-41159-5_7

into different components, one of which is the additive component. Such partition is called a model of genetic effects or a genotype-to-phenotype (GP) map. When the effects of different environmental conditions are also considered, it could be more precisely called a model of genetic and environmental effects or a genotype-and-environment-to-phenotype map—but for simplicity we shall continue to refer to it as a GP map.

Fisher's (1918, 1930) developments had to be implemented significantly over time for several reasons. For instance, Fisher (1918) took the position of disregarding epistasis to be implemented in detail in his theory, a view Wright would argue profusely against (Provine 1986). Kempthorne (1954) and Cockerham (1954) eventually provided developments to integrate epistasis in quantitative genetics models. Later on, these models had to be deeply revised in order to adapt them to find the genetic basis of traits in gene-mapping experiments (e.g. Cheverud and Routman 1995; Hansen and Wagner 2001; Yang 2004; Zeng et al. 2005; Álvarez-Castro and Carlborg 2007).

Within the three previous chapters, we have compiled a wide-ranging and unified theory of genetic effects—a comprehensive GP map. Particularly in what regards the population level, we have provided expressions enabling decompositions of the genotypic values for potentially highly complex genetic architectures at populations with arbitrary departures from equilibrium frequencies, which comprises a very general instalment of the core theory of quantitative genetics. In this chapter, we go beyond the kernel of that mathematical machinery in several ways.

First, we address the decomposition of the genetic variance, which stems from the decomposition of the genotypic values and provides indexes useful to make predictions at the level of the phenotype in populations where relatedness is tracked. The decomposition of the genetic variance makes it particularly evident, from the biological perspective, why orthogonality is necessary as a mathematical property of the models of genetic effects at the population level (whereas not sufficient for the model to be biologically satisfactory). Indeed, such decomposition, many times called partition instead, cuts the genetic variance cake into variances of the different components (the additive and the interaction components) in a way that each slice of the cake—i.e. each component of the variance—reflects the contribution of each component of the model to the total genetic variance. Were the genetic effects non-orthogonal, the decomposition would additionally include non-nil covariances between the different components, thus distorting the aforementioned, desired biological meaning of the variance components—as the non-overlapping slices into which a whole cake has been cut.

We initiate our analysis on this issue by presenting a handy expression to obtain the decomposition of the genetic variance from the genetic-effects design matrix of a genetic architecture at a population (originally provided by Álvarez-Castro and Yang 2011). We also delve into the biological meaning of the decomposition of the genetic variance (particularly following Álvarez-Castro and Le Rouzic 2015). This is a crucial issue in quantitative and evolutionary genetics not only because it burst the development of a vast range of innovative practical applications since when it was first published by Fisher (1918) but also because the details of its correct use and interpretation remains polemic nowadays (see e.g. Hansen 2015).

We next describe a regression method, developed by Nettelblad et al. (2012), that makes the most of the properties of NOIA for obtaining estimates of genetic effects in mapping experiments. This way, we make it sure that the effort of developing such a general theory of genetic effects does not fall on deaf ears in what regards gene mapping. Indeed, it is crucial to apply orthogonal models also when estimating genetic effects from real data in mapping experiments. Using a mathematical model for estimating the genetic effects that fit an experimental sample not only ensures that the estimates obtained make real sense biologically speaking but also that the model selection procedures—necessary to efficiently select the genetic basis of the trait that best fits the data available—can be carried out in the most appropriate mathematical way.

Then, we show that the theory of genetic effects provided in the previous chapters may be further generalized to include additional genetic parameters of contrasted importance—Fisher's (1941) average excesses. Precisely, we present Álvarez-Castro and Yang's (2012) model embracing both average effects and average excesses. We also comment on the extension of that model to more general settings. Orthogonality is maintained in all generalizations.

Adding up this chapter to the previous ones, a fairly general theory of genetic and environmental effects—a general GP map—is here provided, particularly targeted to the demands and possibilities raised at the mapping era. Furthermore, along the current chapter, orthogonality is analysed from different perspectives. Indeed, we end up this chapter with a general view on the role of orthogonality in models of genetic effects.

7.2 Orthogonal Decomposition of the Genetic Variance

The components of the genetic variance (Fisher 1918) provided key concepts for the understanding of how Mendelian factors make individuals to resemble each other in a population as a result of their degree of relatedness. Thus, the decomposition of the genetic variance unlocked the development of procedures to predict the behaviour of quantitative traits in populations along generations. In particular, it became possible to compare the costs of different selection schemes and their expected returns in terms of average phenotype improvement, a cornerstone of animal and plant breeding.

However, it turned unrealistic to obtain accurate estimates of epistatic components of the genetic variance from data on phenotypes and relatedness. Therefore, it is understandable that the motivation of stepping beyond simplistic models of genetic variance decomposition was not very high during several decades. Nowadays, variance decomposition may be viewed from a different perspective. One century of research and, particularly, the knowledge acquired with mapping experiments make it more meaningful to inspect the properties of complex genetic architectures. NOIA (and particularly its upgraded version, ARNOIA) enables us to express in an accurate manner the genetic variance decompositions of complex genetic architectures at populations with arbitrary departures from equilibrium frequencies. A procedure facilitating such task follows.

7.2.1 The Decomposition of the Genetic Variance with NOIA

The COIA regression framework (Expressions 6.1–6.17) leads both to a decomposition of the genotypic values and to a decomposition of the genetic variance, beyond correlations between/among factors (loci and environments). Indeed, Expressions 6.1–6.3, 6.10, 6.14 and 6.16 provide decompositions of the genotypic values that are achieved by following the indications of the remaining expressions of Chap. 6. From each of those cases, the decomposition of the genetic variance can be obtained by computing weighted (by the genotypic frequencies) averages of components across genotypes.

In Chap. 6, we used COIA to develop ARNOIA as an extension of NOIA to non-random associations between/among factors (genes and/or environments). Thus, ARNOIA consists in more general expressions of the kind $\mathbf{G} = \mathbf{SE}$ (equivalently, $\mathbf{E} = \mathbf{S}^{-1}\mathbf{G}$). The $\mathbf{G} = \mathbf{SE}$ expression is a decomposition of the genotypic values. In the particular case of one biallelic locus (Expression 6.1), the $\mathbf{G} = \mathbf{SE}$ expression provides a decomposition of the genotypic values into additive and dominance components. However, in the case of two biallelic loci, for instance, the $\mathbf{G} = \mathbf{SE}$ expression provides a decomposition into the additive effects (A) of each of the loci, the dominance effects (D) of each of the loci, the additive-by-additive component (AA), the dominance-by-additive component (DA), the additive-by-dominance component (AD) and the dominance-by-dominance component (DD). Thus, some additional sums often have to be performed to attain the standard decomposition, with one common additive component and one common dominance component.

As noted by Álvarez-Castro and Yang (2011), one way of transforming the $\mathbf{G} = \mathbf{SE}$ expression into a matrix providing the standard decomposition is:

$$\mathbf{G}_{dec} = \mathbf{S}Diag(\mathbf{E})\mathbf{H}, \tag{7.1}$$

where *Diag* operates on vectors and provides a diagonal matrix with the vector in the diagonal and $\mathbf{H}$ is a matrix operator to sum the columns of the parameters related to a common component (as further explained below). Each row of $\mathbf{G}_{dec}$ is built up with the components in which the corresponding genotypic value has been decomposed—the first row gathers the summands in which the first genotypic value is decomposed and so on.

In the case of two biallelic loci, we have to sum the columns of the additive effects of the two loci—the second and fourth columns—and the columns of the dominance effects of the two loci, the third and the seventh columns. The required operator $\mathbf{H}$ in Expression 7.1 for this case is:

$$\mathbf{H} = \begin{pmatrix} 1 & 0 & 0 & 0 & 0 & 0 & 0 \\ 0 & 1 & 0 & 0 & 0 & 0 & 0 \\ 0 & 0 & 1 & 0 & 0 & 0 & 0 \\ 0 & 1 & 0 & 0 & 0 & 0 & 0 \\ 0 & 0 & 0 & 1 & 0 & 0 & 0 \\ 0 & 0 & 0 & 0 & 1 & 0 & 0 \\ 0 & 0 & 1 & 0 & 0 & 0 & 0 \\ 0 & 0 & 0 & 0 & 0 & 1 & 0 \\ 0 & 0 & 0 & 0 & 0 & 0 & 1 \end{pmatrix}. \tag{7.2}$$

If a decomposition merging the dominance-by-additive and the additive-by-dominance components into one is preferred, then the penultimate and the antepenultimate columns of matrix **H** in Expression 7.2 should be merged into one—with ones in the sixth and the eighth positions and zeros otherwise. For merging all epistatic components into one, as in Expression 6.2, the four last columns of matrix **H** in Expression 7.2 should be merged into one column with ones in the fifth, sixth, eighth and ninth positions and zeros otherwise. Incidentally, Expression 6.2 would not provide results enough to build up a NOIA (or ARNOIA) model, as Expression 6.3 does.

From Expression 7.1—with any of the choices of matrix **H** commented right above—it is straightforward to obtain a decomposition of the genetic variance, simply as (Álvarez-Castro and Yang 2011):

$$\mathbf{V} = \mathbf{P}_G(\mathbf{G}_{dec} \circ \mathbf{G}_{dec}), \tag{7.3}$$

where $\circ$ stands for the Hadamard product of arrays (providing the array with the products of the elements at the same position of the two arrays) and $\mathbf{P}_G$ is the row-vector of genotypic frequencies, $\mathbf{P}_G = (p_{ijkl})$. The resulting vector, $\mathbf{V}$, is a row-vector with the squared mean at the first position and the different variance components (the actual decomposition of the genetic variance) at the remaining positions, $\mathbf{V} = (\mu^2, V_\mathrm{A}, V_\mathrm{D}, V_\mathrm{AA}, V_\mathrm{DA}, V_\mathrm{AD}, V_\mathrm{DD})$. The variance components in $\mathbf{V}$ may differ from these depending upon the choice of **H** made, as explained above.

7.2.2 Definition and Meaning of the Variance Decomposition

The decomposition of the genetic variance expresses the genetic variance, V_G, as a sum of terms accounting for the relative weight of different components into which the genetic composition of the population has been split. Let us recall on the one hand what the genetic variance is and, on the other hand, the criterion applied to choose the particular components into which it is split. The genetic variance, V_G, is the variance of the genotypic values (see Expression 2.3 and related text) and it is one of the terms into which the (total) phenotypic variance, V_P, is split. The other term is the (random) environmental variance, V_E (see Expression 2.20). These two quantities account for the relative weights of the genetic determination of the trait and the (random) environmental contribution, respectively (as already exposed in Expression 2.21 and related text).

The variance component of genetic origin, V_G, is further split. We start by considering one only locus, in which case it becomes evident that how much (if at all) the heterozygotes depart from the midpoint between the corresponding homozygotes—i.e. the degree of dominance—determines the properties of the system (see Fig. 2.1). Indeed, the outcome of selection on the trait may wobble from fast fixation of one predetermined allele to fixation of alternative alleles for different initial frequencies in the population to maintenance of genetic variability, depending upon the degree of dominance. This is the reason why the genetic component of a one-locus case, G, is split by detaching a dominance component, D, from it (Fisher 1918). The resulting decomposition consists of one component accounting for the difference between the homozygotes—the additive component, A—and another component accounting for the (putative) departure of the heterozygotes from the midpoint of the two corresponding homozygotes, the dominance component, D.

The split of the genetic component into additive and dominance components, $G = A + D$, raises the concept of interaction. Indeed, the additive component accounts for genetic effects of allele substitutions that add up whereas the dominance component accounts for departures from the additive behaviour that arise when the effects of different alleles come together and may, thus, interact (as more in detail discussed in Chap. 2). As discussed in the initial chapters of this book (Chaps. 1 and 2), dominance was crucial already for Mendel to derive his interpretation of the proportions he found, and thus it was a mainstream concept from the origins of the field of genetics. In any case, the decomposition of the genetic variance, V_G, into additive, V_A and dominance, V_D, components just comes as consequence of the detachment of the dominance component at the level of the genetic effects.

Thus, we may summarize that, by definition, the meaning of the decomposition of the phenotypic variance, V_P, is as follows. The additive component of the genetic variance, V_A, accounts for the proportion of the genetic variance in a population that can be explained in terms of the additive effects of allele substitutions. The dominance component, V_D, accounts for the variance caused by within-locus interactions. These two components constitute the genetic variance, V_G, which accounts for the variation in phenotype in the population due to the genetic determination of the individuals. Lastly, the environmental component, V_E, accounts for the variation in phenotype in the population caused by the (random) environment on the individuals. Thus, we can write (see Expressions 2.20, 2.23 and 2.24):

$$V_P = V_G + V_E = V_A + V_D + V_E. \tag{7.4}$$

With several loci, more interactions are possible—epistatic interactions. Thus, additional components of the genetic variance also spring (see Expressions 2.33, 2.35–2.37). These emerging components imply combinations of additive effects and dominance effects (e.g. V_{AA}, V_{DA}, V_{AD}, V_{DD}, V_{AAA}, V_{DAA},. . .). In any case, all of them keep on accounting for the proportion of (total) phenotypic variance caused by the corresponding genetic interactions in the individuals of the population.

7.2.3 Different Perspectives and Essential Reminders

The decomposition of the genetic variance enables to describe the resemblance between relatives (the correlation of the measurements of the trait in pairs of related individuals) in the population (see Expressions 2.26–2.30). As expected from everyday experience, the closer the relatedness, the higher the average resemblance. Amongst all kinship cases, the resemblance between parents and offspring is key to analyse the outcome of selection on the phenotype and, thus, to ponder the payoff of an investment in a selection program. It happens that of all the observed trait variance, V_P, that due to components implying only additive effects can be harnessed by selection on the phenotype.

The (narrow-sense) heritability, h^2 (defined in Expression 2.25 as $h^2 = V_A / V_P$) is the index used to predict the response to selection (see Expression 2.28 and related text). For this approach to work, it is necessary to estimate the heritability from phenotypes and relatedness. This procedure is subject to (often ignored) intricacies. This was noted in the final section of Chap. 2, and it is not going to be resumed here (for additional discussion on this issue see e.g. Houle 1992; Hansen et al. 2011).

This is the mainstream perspective on the decomposition of the genetic variance that stems from the roots of quantitative genetics. It naturally generated a strong inertia that persists to this day. Nonetheless, additional approaches based on the decomposition of the genetic variance are possible, as genetic architectures of experimental (and even natural) populations have been inspected for already three decades (see e.g. Rifkin 2012). Thus, it makes sense to use the decomposition of the genetic variance to analyse the properties of genetic architectures—that may have been disclosed or can be deemed realistic with the results of mapping experiments in mind—under different population compositions. This perspective catapults the epistasis components onto the research agenda, since they were unrealistic to be estimated from phenotypes and relatedness. With that in mind, we hereafter summarize a few crucial considerations about the meaning and implications of the decomposition of the genetic variance that are more often than desired overlooked (see e.g. Álvarez-Castro and Le Rouzic 2015 for more detailed explanations).

First, strictly speaking, the aforementioned heritability approach is designed to predict the outcome of selection in a one-generation step. It is sensible to infer that extrapolations of those predictions to a longer-term selection process may work to some extent. However, mid- and long-term predictions based on the estimation of heritability at one generation are not supported theoretically.

Second, the interaction components inform about the extent to which those components are affecting the selection process with a particular population composition (i.e. with particular population frequencies). Low or even negligible values of interaction components of the genetic variance do neither imply that the corresponding interactions are absent nor that they will not be important in the selection process in forthcoming generations. Instead, it simply means that they are not affecting selection in the next generation step.

Third, low additive variances and low heritabilities do not necessarily imply lack of genetic variability. When either the selection process starts close to an unstable equilibrium (i.e. a minimum) or it goes through the neighbourhood of a saddle point, the heritability may be very low in the face of high genetic variability. That variability may get decanalized after a number of generations with very low selection response. There also is the possibility of maintenance of genetic variability at a stable equilibrium (i.e. a fitness maximum). Indeed, not only fixation points are stable equilibria—stable polymorphic equilibria exist just as well. If this occurs, all genetic variance is in the form of dominance variance, which holds even when strong epistasis is involved. These latest cases shall be treated in more detail in the next chapter.

7.2.4 Two Additional Remarks

For simplicity, we have so far considered only random environmental effects. In the previous chapters (Chaps. 4–6), we have considered particular environmental conditions that can be monitored. In Chap. 6, for instance, we have mentioned that the individuals of the population could develop under three different average temperatures. Such situations lead to decompositions with additional components—one term for the monitored environments and gene-environment interaction terms (see e.g. Expression 6.10). In what regards the variance decomposition, on top of those emerging interaction components, it may be convenient to keep track of the purely genetic components separately for each environment. As explained above, the decision on which components to merge or not is made effective in the theoretical model through the choice of matrix **H** (see Expressions 7.1 and 7.2 and related text).

Throughout this and the previous chapters of this book, we have considered models of genetic effects as GP maps (whether or not environmental conditions are monitored). Focusing on the variance decomposition makes it particularly evident that these models enable the analysis of selection on the trait under study. Implicitly, we have assumed the action of directional selection. It is to be kept in mind that the variance decomposition of a trait is not informative at all about the action of stabilizing selection on it. In other words, the models in this book are completely useful when they are genotype-to-fitness (GF) maps (including within this category also genotype-and-environment-to-fitness maps). Indeed, by definition, there is directional selection on fitness. Since breeding is about improving traits, directional selection is over and over again assumed throughout quantitative genetics.

7.3 Orthogonal Estimation of Genetic Effects from Real Data

As pointed out in the previous section, orthogonality is a mathematical property that a model of genetic effects fulfils as a consequence of its parameters fitting their biological meaning. Thus, it is necessary to express the genotypic values (and the genetic variance) in terms of components that have to be orthogonal in order to acquire their correct meaning, which does not mean that orthogonality by itself suffices for a decomposition to be the correct one. Therefore, for estimating genetic effects from real data with the desired biological meaning, the model used necessarily has to be orthogonal. Besides, in order to ensure that the resulting estimates are orthogonal and fit to the desired biological meaning, it is also necessary to make the correct choice of estimation procedure. In this section, we present and discuss Nettelblad et al. (2012) proposal for obtaining appropriate (orthogonal) estimates of genetic effects in the face of missing data.

7.3.1 Estimates of Genetic Effects Take Priority

First, we review the basics of estimation of genetic effects. The data consists of a number of individuals that have been phenotyped and genotyped at a number of markers. For simplicity, we consider one only biallelic marker, M, with genotypes M_1M_1, M_1M_2 and M_2M_2. The extension to more complex genetic architectures is straightforward—it just generates matrices with higher dimensions. Let $\mathbf{Y}$ be the column vector of phenotypes of the N individuals of our experimental population and $\mathbf{Z}$ the assignment-matrix, assigning each individual to its marker genotype, with as many rows as individuals, N, and as many columns as genotypes—in our case, three. In particular, the nth row of $\mathbf{Z}$ has a one at the position of the marker genotype of the nth individual (in the first position if the genotype is M_1M_1, in the second position if it is M_1M_2, in the third position if it is M_2M_2) and zeros otherwise. The genotypic values could then be estimated through the regression:

$$\mathbf{Y} = \mathbf{Z}\mathbf{G} + \boldsymbol{\eta}, \tag{7.5}$$

where $\boldsymbol{\eta}$ gathers the error terms.

From genotypic values so obtained, additive and dominance values could be derived using the matrix notation as $\mathbf{E} = \mathbf{S}^{-1}\mathbf{G}$ (Expression 5.11, with the population frequencies of the data). However, as explained in the previous section, the evolutionary importance of the genetic effects is primordial. Indeed, for pondering the effect of a locus on the phenotype, we do not only want to have estimates of the genotypic values but to test whether they are different from each other and obtaining estimates of the genetic effects and their errors enable us to easily perform such test. Since the genetic effects are themselves of crucial evolutionary importance, as mentioned right above, they are indeed the best possible way of performing that test.

As a consequence of that, if we obtain an estimate of a genetic effect that is not statistically different from zero, we shall typically assume that it equals zero. Then, in order to be coherent with this, the genotypic values, $\mathbf{G}$, must be computed from the genetic effects that have been found to achieve statistical significance, $\mathbf{E}$. This can be done by just applying these genetic effects to the alternative expression in matrix notation $\mathbf{G} = \mathbf{S}\mathbf{E}$ (Expression 5.12). This correction of the genotypic values by the selected genetic effects is called genetic filtering (for further discussion see Álvarez-Castro 2012).

Thus, the correct genotypic values to work with will typically be different from the ones that would be obtained with Expression 7.5. And, beyond that, it could even be unfeasible in practice to obtain all genotypic values using that expression, particularly when complex genetic architectures are to be considered. For instance, as pointed out by Álvarez-Castro (2012, but for a misprint here corrected) "[w]ith a network of just 4 loci and two alleles per locus, the number of individuals of each of the [16] possible complete homozytotes in an F_2 population of 800 individuals is just 3.125—hence, some empty genotypic classes will most surely occur".

Overall, the meaningful (or even just obtainable) genetic values in a gene-mapping study will be those resulting from a genetic filtering of the genetic effects. Fortunately, Expression 7.5 can be further worked out to provide the genetic effects (and, thus, their statistical significance) directly from the data. To do so we just need a model of genetic effects, i.e. an expression transforming the genotypic values into genetic effects—if one prefers, a decomposition of the genotypic values into genetic effects. The convenient matrix notation to which we have been systematically adhering in the previous chapters is adequate to facilitate this task. Indeed, the aforementioned expression $\mathbf{G} = \mathbf{SE}$ makes it straightforward to transform Expression 7.5 into:

$$\mathbf{Y} = \mathbf{ZSE} + \boldsymbol{\eta}. \tag{7.6}$$

The particular expression $\mathbf{G} = \mathbf{SE}$ to use is, for our current case, the NOIA model for one biallelic locus at the population level (Expression 5.12), implemented with the marker genotype frequencies.

The estimates obtained with Expression 7.6 are orthogonal because the design matrix in that expression, $\mathbf{X} = \mathbf{ZS}$, is orthogonal (see Nettelblad et al. 2012 for details). Now, it is noteworthy that a key fact of the assignment-matrix $\mathbf{Z}$ for the design-matrix $\mathbf{X}$ to inherit the orthogonality of the genetic-effects design matrix $\mathbf{S}$ is that it has one only non-nil value per row. This is particularly important when it finally comes to getting the researcher's hands dirty with the common hazards of real data, as explained hereafter.

7.3.2 Implementing Genotype Probabilities

In real cases the researcher will not be able to definitely assign each individual to a genotype. To start with, real experiments suffer from incomplete genotypic data. Besides, in the case of quantitative-trait-locus (QTL) analyses, inspecting regions between flanking markers is required. In both cases, probabilities for each individual to be assigned to a genotype are computed, in the absence of a certain assignment. The most evident way of implementing such genotype probabilities in matrix $\mathbf{Z}$ is to just fill each row of $\mathbf{Z}$ with the genotype probabilities of the corresponding individual, instead of a certain assignment based on a one and zeros, a method called Halley-Knott regression (HKR, Haley and Knott 1992).

However, HKR leads to an assignment matrix, $\mathbf{Z}$, whose rows no longer have one only non-nil value per row and, therefore, orthogonal estimates are no longer warranted. As explained above, this is a crucial issue, as orthogonality is not only a convenient mathematical property but, most of all, one of the necessary conditions for obtaining meaningful estimates. With this in mind, it is not surprising that Nettelblad et al. (2012) further showed that HKR provides incongruous estimates also at the level of the genotypic values. Indeed, HKR may lead to genotypic values even lying outside the range of all phenotypes observed in the experiment (Nettelblad et al. 2012).

Opportunely, Nettelblad et al. (2012) proved that the aforementioned problems can be overcome by developing a procedure that applies regression imputation—interval mapping by imputations (IMI). In this procedure, each individual is represented by as many rows in the assignment-matrix, **Z**, as possible genotypes. Thus, the number of rows comes to be the one of the assignment-matrices considered right above (i.e. number of individuals, N) times the number of genotypes—which in our current case is three. Incidentally, the number of columns remains as above, i.e. as many as genotypes (again, three).

In IMI, the increase in dimension of the assignment matrix serves to accommodate each (putatively non-nil) genotype probability into one single row. To be precise, the values entering the IMI assignment matrix are not the genotype frequencies themselves, but their square roots. In particular, for each individual, each genotype probability square root enters the assignment matrix at its row and at the column corresponding to the genotype in question, the remaining scalars of that row being zeros. Finally, for applying Expression 7.6 using an IMI assignment-matrix, **Z**, it is just necessary to modify the vector of phenotypes, **Y**, so that each individual's phenotype is repeated as many times as possible genotypes (in our case, three), thus getting to a column vector with the same scalars as **Z** has rows.

As Nettelblad et al. (2012) showed, the assignment matrix of IMI can be built by multiplying a diagonal matrix, $\check{\mathbf{D}}$, with the square roots of the genotype probabilities of the individuals and a matrix built as a column of as many blocks as individuals, each block being an identity matrix of dimension equal to the number of genotypes, that is:

$$\mathbf{Z} = Diag\begin{pmatrix} \sqrt{p_{11}^{1}} \\ \sqrt{p_{12}^{1}} \\ \sqrt{p_{22}^{1}} \\ \hline \sqrt{p_{11}^{2}} \\ \sqrt{p_{12}^{2}} \\ \sqrt{p_{22}^{2}} \\ \hline \vdots \\ \hline \sqrt{p_{11}^{n}} \\ \sqrt{p_{12}^{n}} \\ \sqrt{p_{22}^{n}} \end{pmatrix} \begin{pmatrix} 1 & 0 & 0 \\ 0 & 1 & 0 \\ 0 & 0 & 1 \\ \hline 1 & 0 & 0 \\ 0 & 1 & 0 \\ 0 & 0 & 1 \\ \hline & \vdots & \\ \hline 1 & 0 & 0 \\ 0 & 1 & 0 \\ 0 & 0 & 1 \end{pmatrix}, \tag{7.7}$$

where p_{ij}^{n} is the probability of individual n to be of genotype $MiMj$. The horizontal lines in Expression 7.7 just make it easier to visualize the blocks accounting for the information of each of the different individuals.

For applying Expression 7.6 with the IMI assignment matrix in Expression 7.7 it is necessary to enlarge the vector of phenotypes accordingly (each phenotypic observation must appear as many times as possible genotypes, i.e. three in the present case) and to multiply it with the diagonal matrix, $\check{\mathbf{D}}$. This provides estimates of genetic effects with the desired meaning (and, thus, in particular, orthogonal) in the face of missing genotype data. Besides, it provides congruent estimates of genotypic values. As mentioned above, the extension of this procedure to more complex genetic architectures does not significantly increase the conceptual complexity of Expression 7.7, although it could (highly) increase the dimension of the matrices implied.

7.4 An Orthogonal Framework Also Entailing Average Excesses

Fisher (1918) defined the additive effect, α, as the average effect of an allele substitution in a population (see Chaps. 2, 5 and 6). With this, he established a theoretical framework describing the phenotypic resemblance between relatives, which enabled predictions about phenotype change under selection using data on relatedness and phenotype. Later on, Fisher (1930, 1941) proposed another genetic parameter useful to understand the evolutionary properties of a population, the average excess, α^{*}. Hereafter, we shall extend NOIA to also accommodate the average excess concept. In other words, the first component into which NOIA decomposes the genotypic values (and the genetic variance) will not necessarily be related to the average effects of allele substitutions (as we have conceived it hitherto), but it will be flexible enough as to account for the average excess concept as well.

7.4.1 Extension of NOIA to Entail Average Excesses

To start with, we consider one biallelic locus, A, with alleles A_1 and A_2, and propose the following modification of Kempthorne's (1957) regression framework $\mathbf{G} = \mathbf{1}\,\mu + c\mathbf{N}\bar{\boldsymbol{\alpha}} + \bar{\boldsymbol{\delta}}$, expanding to:

$$\begin{pmatrix} G_{11} \\ G_{12} \\ G_{22} \end{pmatrix} = \begin{pmatrix} 1 \\ 1 \\ 1 \end{pmatrix} \mu + c \begin{pmatrix} 2 & 0 \\ 1 & 1 \\ 0 & 2 \end{pmatrix} \begin{pmatrix} \bar{\alpha}_1 \\ \bar{\alpha}_2 \end{pmatrix} + \begin{pmatrix} \bar{\delta}_{11} \\ \bar{\delta}_{12} \\ \bar{\delta}_{22} \end{pmatrix}. \tag{7.8}$$

The only difference between this regression framework and the one proposed by Kempthorne (1957), which we used to obtain NOIA (Expression 5.6), is that here we added a parameter c. In other words, NOIA is a particular case of Expression 7.8

with $c = 1$, thus using the design-matrix **N** as a particular case of c**N**. Since Expression 7.8 entails more possibilities than Expression 5.6, it uses a slightly different notation for the explanatory variables and for the error terms—the additive and the dominance components, $\bar{\boldsymbol{\alpha}}$ and $\bar{\boldsymbol{\delta}}$, respectively.

Let p_i be the frequency of allele A_i and p_{ij} be the frequency of genotype A_iA_j. Analogous to the derivation of the formulation of the kind $\mathbf{E} = \mathbf{S}^{-1}\mathbf{G}$ at the population level (Expression 5.11) from the basic regression framework of NOIA (Expression 5.6), from Expression 7.8, it follows:

$$\begin{pmatrix} \mu \\ \bar{\alpha} \\ \bar{\delta} \end{pmatrix} = \begin{pmatrix} p_{11} & p_{12} & p_{22} \\ -\dfrac{p_{11}p_2}{c(2p_1p_2 - \frac{1}{2}p_{12})} & \dfrac{p_{12}(p_1 - p_2)}{c(4p_1p_2 - p_{12})} & \dfrac{p_1p_{22}}{c(2p_1p_2 - \frac{1}{2}p_{12})} \\ -\frac{1}{2} & 1 & -\frac{1}{2} \end{pmatrix} \begin{pmatrix} G_{11} \\ G_{12} \\ G_{22} \end{pmatrix}. \tag{7.9}$$

Reassuringly, this expression leads to the $\mathbf{G} = \mathbf{SE}$ formulation of NOIA at the population level (Expression 5.11) as a particular case, with $c = 1$. Then, by inverting matrix $\mathbf{S}^{-1}$ in Expression 7.9, we obtain the equivalent formulation of the kind $\mathbf{G} = \mathbf{SE}$, expanding to:

$$\begin{pmatrix} G_{11} \\ G_{12} \\ G_{22} \end{pmatrix} = \begin{pmatrix} 1 & -2cp_2 & -\dfrac{p_{12}p_{22}}{2p_1p_2 - \frac{1}{2}p_{12}} \\ 1 & c(p_1 - p_2) & \dfrac{p_{11}p_{22}}{p_1p_2 - \frac{1}{4}p_{12}} \\ 1 & 2cp_1 & -\dfrac{p_{11}p_{12}}{2p_1p_2 - \frac{1}{2}p_{12}} \end{pmatrix} \begin{pmatrix} \mu \\ \bar{\alpha} \\ \bar{\delta} \end{pmatrix}. \tag{7.10}$$

Once again, this reduces to the $\mathbf{G} = \mathbf{SE}$ formulation of NOIA (Expression 5.12) with $c = 1$.

Expression 7.10 has also been derived by Álvarez-Castro and Yang (2012), using an alternative procedure—by building up statistical contrasts, an approach also followed by e.g. Cockerham (1954) and Zeng et al. (2005) for obtaining other models of genetic effects. Álvarez-Castro and Yang (2012) showed that beyond providing the average effects (i.e. $\bar{\boldsymbol{\alpha}} = \boldsymbol{\alpha}$) with $c = 1$, Expression 7.10 also provides the average excesses (i.e. $\bar{\boldsymbol{\alpha}} = \boldsymbol{\alpha}^*$) with $c = 1/(1 + F)$, where $F = 1 - p_{12}/2p_1p_2$ is Wright's (1965) fixation index. The average effects, $\boldsymbol{\alpha}$, and the average excesses, $\boldsymbol{\alpha}^*$, coincide when $F = 0$, i.e. under Hardy-Weinberg proportions (see Table 7.1, taken from Álvarez-Castro and Yang 2012).

7.4.2 Effective Gene Content

More to the point, Álvarez-Castro and Yang (2012) identified the parameter c as the effective gene content, which enables "to capture the correlation between alleles that

Table 7.1 Summary of some relevant mathematical and biological features associated to different statuses of the heterozygosity of a population. Taken from Álvarez-Castro and Yang (2012)

Heterozygotes deficiency	Observed heterozygotes fit HWE	Heterozygotes excess
$0 < F \leq 1$	$F = 0$	$-1 \leq F < 0$
$½ \leq c < 1$	$c = 1$	$c > 1$
$\lvert\alpha^*\rvert > \lvert\alpha\rvert$	$\alpha^* = \alpha$	$\lvert\alpha^*\rvert < \lvert\alpha\rvert$
Assortative mating or homozygotes favoured or population structure	Random mating and either no selection or geometric fitnesses	Dissassortative mating or heterozygotes favoured or gene duplication

the average excesses account for". In other words, the effective gene content makes the apparent content of the alleles at the gene to differ from the actual one due to departures from the Hardy-Weinberg proportions. Figure 7.1 shows the graphical interpretation of the gene content provided by Álvarez-Castro and Yang (2012) (cf. Fig. 5.1).

Indeed, as the Hardy-Weinberg proportions occur with the alleles being randomly associated in the individuals, departures from those frequencies imply non-random associations, i.e. correlations between the alleles. As Table 7.1 shows, positive fixation indexes and gene contents smaller than one occur when the alleles are positively correlated (deficiency of heterocygotes), whereas negative correlations (excess of heterozygotes) lead to negative fixation indexes and allele contents greater than one. Fisher (1941) noted that these correlations implied an effective additive contribution of the alleles in the population different from that under Hardy-Weinberg proportions, and the graphical interpretation (Fig. 7.1) shows that the differences in effective contribution can be explained by different effective allele contents. In particular, Fig. 7.1 shows that the effective additive contribution coming from a correlation measured in terms of the fixation index as $F = -2/5$ implies a smoother regression slope, which can be explained as the result of an effective allele content of $c = 5/3$).

7.4.3 More Complex Genetic Architectures at Hand

The extension of Expressions 7.9 and 7.10 to an arbitrary number of biallelic loci in linkage equilibrium is straightforward, using the Kronecker product as in previous models (see e.g. Expression 3.3). Departures from linkage equilibrium can be implemented as well, since the extension of regression framework in Expression 7.8 can be done in the same way as NOIA was upgraded to ARNOIA in the previous chapter. To end up with, we comment on how to further implement this theory to multiple alleles.

The expression of the effective gene content of the average excesses, $c = 1/(1 + F)$, can be retrieved from the expression relating the average effects and the average excesses, $\alpha_i = \alpha_i^* /(1 + F)$, which also applies to multiple alleles

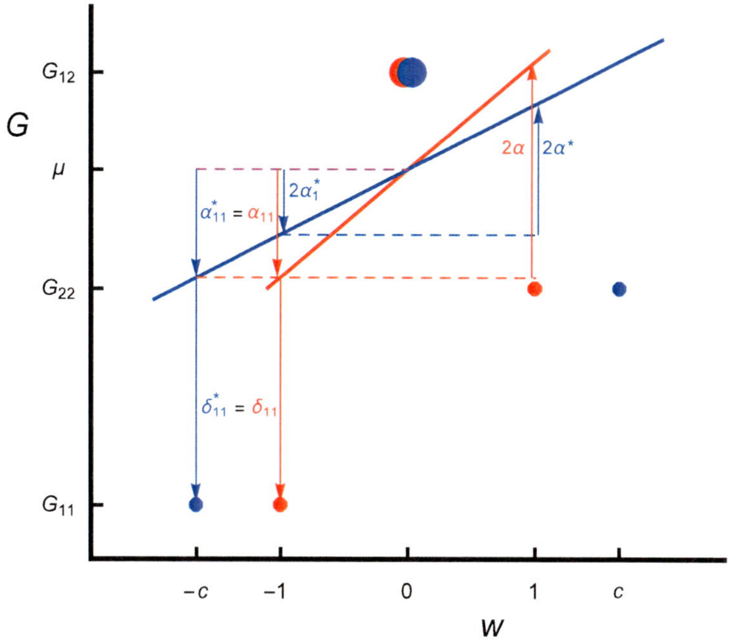

Fig. 7.1 From Álvarez-Castro and Yang (2012). Graphical interpretation of the decomposition of the genotypic values (Expression 7.10) through the statistical excess (in blue) and the statistical (in red) formulations of NOIA for one locus with two alleles. For simplicity, a case with equal allele frequencies ($p1 = p2 = ½$) is shown. The specific genotypic values (circles; $G11 = 1$, $G12 = 3$, $G22 = 2$) displaying overdominance and a fixation index of $F = -2/5$ have been chosen for facilitating the visualization of the parameters of interest. The size of the circles represents the frequency of the genotypes. Horizontal dashed lines emphasize coincident arrow edges, the upper one corresponding to the population mean phenotype, $\mu = 2.55$. The regression independent variable of the statistical formulation is the gene content, whereas the one of the statistical excess formulation is scaled by $c = 1/(1 + F) = 5/3$, and it works as an effective gene content. For both cases, the independent variable, w, is rescaled by its expectation

(Kempthorne 1957). This fact provides the key to develop an extension of Expressions 7.8–7.10 to multiple alleles. More precisely, the effective gene contents of each pair of alleles should be implemented in the design-matrix of the additive component of the regression framework so that the regression is performed with the genotypic values at the desired distances. For three alleles, for instance, Expression 5.14 would have to be modified along these lines in order for the distances between the projections of the genotypic values in the horizontal plane to reflect the effective gene contents (see Fig. 5.2a). We shall omit the mathematical details here.

7.5 Orthogonality Under Scrutiny

Along the three cases considered in this chapter, orthogonality has been addressed from different perspectives. We have first come into the details of the decomposition of the genetic variance. Next, we dealt with the minutes of obtaining estimates of genetic effects in mapping experiments. Finally, we considered a general theory of which both the models of genetic effects developed in the previous chapters and the average excesses can be obtained as particular cases. In the remaining of this chapter, we shall delve on orthogonality in models of genetic effects, particularly in the light of the cases considered above.

7.5.1 Orthogonal Decomposition of the Genetic Variance

When analysing the decomposition of the genetic variance, it became clear that no appropriate partition of the variance is possible with a non-orthogonal model, simply because it would not lead to a set of positive terms summing up the total variance. Metaphorically speaking, our partition aims to cut a cake (the genetic variance) into pieces (the variance components) and it would not help that the pieces overlap or that, taken together, they do not comprise the whole cake.

When cutting a cake, we naturally have in mind some conditions the resulting pieces must meet—we may, for example, want all the pieces to be the same size so as not to make any guest feel less welcome. Analogously, the components of the genotypic values and of the genetic variance must not only be proper components (of a proper partition), but they must also adjust to precise biological meanings. In other words, orthogonality is a necessary condition of models of genetic effects, but not a sufficient one.

As expounded repeatedly from Chap. 2, the components of the genotypic values reflect either effects of allele substitutions between genotypes (i.e. at the individual level), or averages of effects of substitutions over the individuals in a population (i.e. at the population level). In what regards its meaning, there is no difference between the genetic variance decomposition presented in Chap. 2 and Expressions 7.1–7.4. What these expressions add up is an extension of the scope of application of the variance decomposition. In particular, they provide exact mathematical descriptions of the decomposition of the genetic variance under all circumstances implemented in the previous chapters in the NOIA model (including its ARNOIA upgrade).

Thus, the components of the genetic variance are defined as the variances of the components of the genotypic values at the population level, either in Chap. 2 or in this chapter. Biologically, they are key to understanding the resemblance between relatives in the population and, thus, the action of selection on the trait at a population. Each variance component is related to the influence of the corresponding component of the genotypic values on the response to selection at the population considered.

Developing the decomposition of the genetic variance and understanding its meaning has brought forth numerous invaluable applications in the field of quantitative genetics for more than one century. It is however crucial to keep in mind that inferring variance components from phenotypes and relatedness at a population is not very informative about the genetic architecture underlying the trait under study. This implies strong limitations on the use of the results thus obtained, particularly with regard to the time frame along which any prediction made at a certain point may be reasonably reliable.

7.5.2 Orthogonal Estimation of Genetic Effects

Disclosing the genetic architecture of a trait would provide a deeper understanding of the selection process, thus enabling reliable predictions on longer time frames. Mapping experiments have been performed for over four decades, having thus given us clues about the genetic architectures of many traits of interest (see e.g. Rifkin 2012). However, it must be considered that appropriate gene mapping should rely on orthogonal estimation of genetic effects. Only in recent publications, including the present book, a general theory of orthogonal genetic effects (NOIA and, particularly, its upgrade to ARNOIA) has been developed. Therefore, this theory should be applied if we are to disclose new genetic architectures and to question already disclosed ones, as further discussed below.

In the context of a mapping experiment, let us consider the case of having to test whether two biallelic loci are influencing the trait. We have the genotypes of the individuals of the experiment at the two loci and their phenotypes. If we use a two-locus epistatic model to estimate the genetic effects and the epistasis estimates are not significant, we should discard epistasis from our model (a procedure known in general terms as model selection). But then there are two options to do so. We can either keep the marginal effects already estimated or we can estimate them again with a non-epistasis model. If the two options provide different estimates, which ones should we choose? Would any of the two be the correct ones?

When non-orthogonal models are used, the above questions can be replied by saying that none of the options provide correct estimates since, as discussed above, orthogonality is a necessary condition for the estimates of genetic effects to fit to their correct biological meaning. When orthogonal models are used, the first question just fades away since the two options provide the same estimates of marginal effects (and, actually, the same errors associated to the estimates). Concerning the second question, the orthogonal estimates are the correct ones not only when the model used is orthogonal, but if it is also the correct one in what regards the biological meaning of the estimates, so are the models provided in the previous chapters of this book.

The same applies to comparing a two-locus model against a one-locus model. If we want to test a two-locus model against the two possible reduced one-locus models (again a model selection procedure), we must evaluate the statistical significance of the estimates of each of the loci. With non-orthogonal models, both the estimates and their errors are different when we use a two-locus epistatic model, a two-locus non epistatic model and a one-locus model, which makes the model selection procedure more difficult—besides, as mentioned above, all of the estimates being actually incorrect. With an orthogonal model, we just have to obtain a putatively reduced model by removing from the (complete) two-locus epistatic model any effects that are not statistically significant, whether they are epistatic effects, dominance effects, additive effects or a whole locus.

Once the model selection procedures have revealed the genetic architecture best fitting the data, the genotypic values, **G**, coherent with the genetic effects of that genetic architecture, **E**, can be easily obtained, as **G** = **SE**. In this process, called genetic filtering, the particular genetic-effects design matrix that has to be used, **S**, must be the one matching the meaning of the genetic effects (Álvarez-Castro 2012). This implies, in particular, that the matrix has to be orthogonal in the context of the data used to obtain the genetic effects. Thus, also for obtaining meaningful genotypic values, orthogonality is crucial.

So far, we have assumed complete genotype data. With missing genotype data, genotype probabilities have to be obtained and implemented and, if an orthogonal model of genetic effects is used, implementing it in IMI (Expressions 7.5–7.7) enables to obtain orthogonal estimates of genetic effects. In fact, Nettelblad et al. (2012) have illustrated with a simple example how IMI prevents the estimates of effects in reduced models to differ from those of the complete model. Thus, in what regards testing whether the epistatic effects are significant, everything works in the same way as described for the above (complete genotype information) case. Of course, this applies also to dominance or any other genetic effect. However, when it comes to testing a two-locus model against the reduced one-locus models, things work differently.

Here it becomes necessary to pay attention to meanings rather than pursuing technical/mathematical properties alone. When removing one locus from a complete two-locus model, orthogonality keeps on warranting that the estimates in the reduced models remain the same. However, we must at this point nuance that we are assuming that we have not changed anything in what we keep in the reduced model. This means in particular that we are assuming that the genotype frequencies do not change. Indeed, with complete genotype frequencies, it just does not make sense to modify the frequencies of a locus when removing another one. But with missing genotype data, we are using information on all the loci in the model for computing the genotype frequencies of each of the loci. Removing a locus from our model means to remove all the influence of that locus over the model and, thus, to compute again the genotype probabilities in the reduced model. Thus, we could say that, with missing genotype data, reducing the number of loci considered in our complete model implies to develop a new smaller model that is not strictly speaking a reduced model of the complete one.

This way, the estimates we obtain after removing a locus are not the same as in the complete model and, then, we are put back to the questions we had previously circumvented. Nevertheless, it is now possible for us to provide an answer since, as mentioned above, our primary objective must be on the meaning of the estimates obtained, rather than to preserve more or less convenient mathematical properties (there is no real gain from easier model selection, for instance, when we are selecting amongst meaningless models). Thus, the correct estimates are those of the reduced model in which we have recomputed the genotype probabilities in the absence of the discarded locus, as opposed to those the estimates kept had in the complete model. Of course, this applies to larger complete models in which may we remove several loci and may keep more than just one in the reduced model.

Overall, with complete genotype data, orthogonal models of genetic effects tremendously facilitate model selection strategies—they enable straightforward ways of testing any genetic architecture against reduced ones. With missing genotype data, orthogonal models of genetic effects and IMI also significantly aid model selection, although missing data itself makes model selection more cumbersome than in the complete-genotype-data case. In any case, we must keep in mind that we do not only want to use orthogonal models and IMI because they make model selection easier. It is appropriate to use them primarily as long as they enable us to obtain estimates that are biologically meaningful. For that, they must be orthogonal, but it would be a mistake to use an orthogonal model facilitating model selection strategies that provides estimates that are not biologically meaningful.

7.5.3 An Alternative Orthogonal Decomposition

In this chapter, we have also shown that NOIA can be considered as one particular case of a set of orthogonal decompositions of the genotypic values for any given population frequencies (Expressions 7.8–7.10). In that set, there is a different orthogonal decomposition of each value the parameter c (the effective gene content) may take. Note that this set is not intended to entail all possible orthogonal decompositions of the genotypic values. In any case, NOIA—which includes average effects of allele substitutions amongst its parameters—emerges with $c = 1$. The interesting point of that set of orthogonal decompositions is that the average excesses emerge also as a particular case of it—with $c = 1/(1 + F)$.

Overall, Expressions 7.8–7.10 illustrate that orthogonal decompositions of the genotypic values with different useful meanings may exist. In other words, not only useless orthogonal models may exist but also useful models that do not exactly fit to the original proposal of models of genetic effects by Fisher (1918). Taken together, the three cases considered in this chapter illustrate orthogonality as a necessary but not sufficient condition for models of genetic effects to be correct, which (particularly at the sight of the last case) also apply to average excesses as useful genetic effects (with a meaning different from that of the average effects of allele substitutions).

References

Álvarez-Castro JM (2012) Current applications of models of genetic effects with interactions across the genome. Curr Genomics 13:163–175
Álvarez-Castro JM, Carlborg Ö (2007) A unified model for functional and statistical epistasis and its application in quantitative trait loci analysis. Genetics 176:1151–1167
Álvarez-Castro JM, Le Rouzic A (2015) On the partitioning of genetic variance with epistasis. In: Moore JH, Williams SM (eds) Epistasis: methods and protocols. Springer, Humana Press, New York, pp 95–114
Álvarez-Castro JM, Yang R-C (2011) Multiallelic models of genetic effects and variance decomposition in non-equilibrium populations. Genetica 139:1119–1134
Álvarez-Castro JM, Yang RC (2012) Clarifying the relationship between average excesses and average effects of allele substitutions. Front Genet 3:30
Cheverud JM, Routman EJ (1995) Epistasis and its contribution to genetic variance components. Genetics 139:1455–1461
Cockerham CC (1954) An extension of the concept of partitioning hereditary variance for analysis of covariances among relatives when epistasis is present. Genetics 39:859–882
Fisher RA (1918) The correlation between relatives on the supposition of Mendelian inheritance. Trans Roy Soc Edinburgh 52:339–433
Fisher RA (1930) The genetical theory of natural selection. Clarendon, Oxford
Fisher RA (1941) Average excess and average effect of a gene substitution. Ann Eugen 11:53–63
Haley CS, Knott SA (1992) A simple regression method for mapping quantitative trait loci in line crosses using flanking markers. Heredity 69:315–324
Hansen TF (2015) Measuring gene interactions. In: Moore JH, Williams SM (eds) Epistasis: methods and protocols. Springer, New York, pp 115–143
Hansen TF, Wagner GP (2001) Modeling genetic architecture: a multilinear theory of gene interaction. Theor Popul Biol 59:61–86
Hansen TF, Pélabon C, Houle D (2011) Heritability is not evolvability. Evol Biol 38:258–277
Houle D (1992) Comparing evolvability and variability of quantitative traits. Genetics 130:195–204
Kempthorne O (1954) The correlation between relatives in a random mating population. Proc R Soc Lond B Biol Sci 143:102–113
Kempthorne O (1957) An introduction to genetic statistics. Wiley, New York
Nettelblad C, Carlborg Ö, Pino-Querido A, Álvarez-Castro JM (2012) Coherent estimates of genetic effects with missing information. Open J Genet 2:8
Provine WB (1986) Sewall wright and evolutionary biology. University of Chicago Press, Chicago
Rifkin S (ed) (2012) Quantitative trait loci (QTL). Springer, New York
Wright S (1965) The interpretation of population structure by F-statistics with special regard to systems of mating. Evolution 19:395–420
Yang RC (2004) Epistasis of quantitative trait loci under different gene action models. Genetics 167:1493–1505
Zeng ZB, Wang T, Zou W (2005) Modeling quantitative trait loci and interpretation of models. Genetics 169:1711–1725

Applied Cases of Advanced Genetic Modelling

8

Abstract

In this chapter, three applications of the theoretical proposals expounded in the previous chapters of this book are presented. The first application addresses the one-locus three-allele polymorphism of the *ACP1* human enzyme in Europe. The genetic variance decomposition is obtained with NOIA and it is used to estimate fitness values explaining the observed genotype frequencies as those of a stable equilibrium. The second case addresses the historical explanation of speciation through Bateson-Dobzhansky-Müller (BDM) incompatibilities. The decomposition of the genetic variance is here obtained with ARNOIA and used to show to what extent departures from both the Hardy-Weinberg equilibrium and linkage equilibrium contribute to hindering the action of selection right after a secondary contact, thus buying extra time for reproductive isolation mechanisms to evolve. The third and last case addresses built-in instances of disease susceptibility under different environmental conditions reflecting situations of importance in the field of precision medicine, and it shows that useful accurate predictions of the coefficient of the disease can be provided using ARNOIA. The three cases illustrate that the implementation of departures from equilibrium frequencies in the mathematical developments presented in this book is key to water the field of evolutionary quantitative genetics.

8.1 Introduction

Conceptual and mathematical developments leading to a comprehensive theory of genetic effects articulated the foundation on which quantitative genetics was supported over one century ago (Provine 1971; Stoltzfus and Cable 2014). In the previous chapters of this book, a general theory of genetic effects has been presented, from its basic concepts to the most advanced models, including theoretical results that had not been previously published. This theory currently embraces an approach

J. M. Álvarez-Castro, *Genes, Environments and Interactions*,
https://doi.org/10.1007/978-3-031-41159-5_8

at the individual level accounting for a wide range of genetic architectures like multiple alleles and both autosomal and sex-linked loci with arbitrary interactions including dominance, epistasis, imprinting, and gene-environment interactions. Also an approach at the population level has here been attained to properly analyze the evolutionary properties of all those genetic architectures at populations with arbitrary departures from equilibrium frequencies, including departures from Hardy-Weinberg equilibrium, from linkage equilibrium, and from random associations between genes and environments.

Such degree of detail would have been of relatively little use before gene mapping sprung in the last decade of the twentieth century (Lander and Botstein 1989). Indeed, it looks reasonable that hitherto the genetics community did not deem it consequential to fine tune the theory of genetic effects. The previous paradigm was naturally limited to working on the supposition of the simplest genetic architectures not blatantly being in conflict with the empirical observations at that time available. Nevertheless, commendable advances were attained during that period by, e.g., Kempthorne (1954, 1957, 1968, 1997), Cockerham (1954) and Van der Veen (1959), as discussed in the previous chapters.

The perception that genetic architectures of traits of interest would gradually be unveiled made researches to face that the theory available did not live up to the emerging challenges. On the one hand, increasingly general models of genetic effects would be required for not imposing limitations on the genetic architectures to be disclosed in the gene-mapping process. On the other hand, more complex models would likely be required also to undertake detailed analyses of the properties of the potentially intrigued genetic architectures expected to be disclosed. Thus, theoretical models of genetic effects gained attention during the early twenty-first century (see, e.g.. Álvarez-Castro and Yang 2015 and references therein).

In the previous chapter, we have touched on the topic of the application of models of genetic effects to gene-mapping. In this chapter, we address the use of models of genetic effects for the analysis of evolutionary properties of several genetic architectures of practical relevance. First, we undertake a multiallelic system under departures from Hardy-Weinberg. In this example, real data is used to obtain fitness estimates from observed genotype frequencies. Next, we delve into a historical proposal to explain the genetics of speciation. In particular, we disclose new evolutionary properties of a two-locus two-allele system under departures from LE. Then, we analyze an index of medical application in a system involving both gene-environment interaction and gene-environment correlation (i.e., departures from random associations between genes and environments). With this, we address concerns raised in the recent demands of adequate quantitative genetics tools in the aid of twenty-first century medical challenges.

Overall, we deal with cases involving multiple alleles under departures from Hardy-Weinberg, epistasis under departures from linkage equilibrium, and gene-environment interaction under gene-environment correlation (i.e., under departures from random associations between genes and environments). This entails a wide spectrum of the theory presented in the previous chapters, while not aiming for a complete coverage of all the theoretical cases considered. In the last section, an overview of the results presented is made to add to the articulation of a further refined evolutionary quantitative genetics.

8.2 The Human *ACP1* Polymorphism in Europe

The extension of NOIA to multiple alleles accounting for arbitrary departures from the Hardy-Weinberg proportions was derived by Álvarez-Castro and Yang (2011), as expounded in Chap. 5. In that paper, the multiallelic NOIA was applied to the study of European variants of the *ACP1* human enzyme, a polymorphism originally discovered by Hopkinson et al. (1963). Three alleles were found to segregate in European populations, *ACP1*A*, *ACP1*B*, and *ACP1*C*, displaying an additive behavior in enzyme activity (e.g., Spencer et al. 1964; Brinkmann et al. 1971; Eze et al. 1974; Greene et al. 2000). Thus, selection favoring high values of enzyme activity would lead to the fixation of the best allele in this regard, *ACP1*A*. The question thus arises about how the polymorphism is maintained.

Based on knowledge about the physiology of the enzyme, an attempt was made to explain the maintenance of the polymorphism in terms of inhibition by folic acid (Sensabaugh and Golden 1978). However, the inhibition results displayed again an additive pattern. Later on, Wilder and Hammer (2004) advocated that allele *ACP1*C* is in its way to disappearance. In particular, they studied allele *ACP1*C* in opposition to the other two alleles as a whole. In other words, they considered a system in which the action of alleles *ACP1*A* and *ACP1*B* is accounted for by one allele cluster, thus being able to apply theory on biallelic models to the study of the three-allele situation in question. Then they compared the observed genotype frequencies of their constructed system with the expected ones under Hardy-Weinberg proportions, and they found the results to support *ACP1*C* to be a "deleterious" allele in the process of being wiped out by selection.

Wilder and Hammer's (2004) analysis illustrates how tempting it can get to be to address the intricacies of complex situations using tools developed with simpler models in mind. But the emergent properties springing in complex models make this kind of approaches inappropriate. The equilibrium properties of multiple alleles systems are a striking example of this (see Fig. 5.2 and related text). Indeed, whereas the superiority of the heterozygote is a necessary and sufficient condition for the maintenance of the polymorphism in a biallelic model, that same condition becomes neither necessary (Kimura 1956) nor sufficient (Mandel 1959) for the maintenance of polymorphisms with more alleles involved.

In general, the high increase in complexity of multiallelic models has been pointed out long ago (Li 1967). This is why it becomes crucial to develop theory that is adequate to the challenges addressed. Lewontin et al. (1978) derived necessary conditions the frequencies have to fulfill for the maintenance of a multiallelic system. Álvarez-Castro and Yang (2011) showed that the observed frequencies of the *ACP1* polymorphism fulfill those conditions. However, since those conditions are not sufficient, this result still leaves open whether the polymorphism is supposed to be maintained or not.

8.2.1 Turning a Verbal Model into Numbers

Greene et al. (2000) elaborated an explanation to the maintenance of the human *ACP1* three-allele polymorphism in Europe through a verbal model. In particular, they proposed that two opposing evolutionary forces conjoin into a stabilizing selection pattern. On the one hand, the highest enzyme activities (particularly genotype *ACP1*C/ACP1*C*) would be penalized for low adaptation to cold temperatures. On the other hand, there is data associating the lowest enzyme activities (particularly genotype *ACP1*A/ACP1*A*) to a higher risk of fetal macrosomia and adult obesity. In particular, a high performance of the heterozygous genotype *ACP1*A/ACP1*B* would prevent *ACP1*A* to be removed by selection. Overall, intermediate enzyme activities would be favored by selection against the extreme ones.

It looks reassuring that empirical observations turn an additive system (which would just lead to the fixation of one allele) into a much more complex genetic architecture, thus leaving the door open to maintenance of genetic variability. However, based on just a verbal model, it is difficult to evaluate whether the arguments provided are sufficient to justify the maintenance of a three-allele polymorphism. It is difficult to ensure, for instance, that the allele most associated to intermediate activity levels, *ACP1*B*, would not end up being brought to fixation, or that at least *ACP1*C* would not be removed by selection—as Wilder and Hammer (2004) considered.

In order to more thoroughly inspect Greene et al. (2000) reasoning, Álvarez-Castro and Yang (2011) turned their verbal model into numerical values, thus being able to perform numerical analyses. In accordance with that verbal model, they set the fitnesses of *ACP1*A/ACP1*B* and *ACP1*C/ACP1*C* to $\omega_{AB} = 1$ and $\omega_{CC} = 0.6$, respectively, and assayed different intermediate values for the remaining fitnesses, ω_{AA}, ω_{BB}, ω_{AC}, and ω_{BC}. They used NOIA to obtain the V_A/V_G ratio for the fitness values considered in populations with the observed frequencies (by Brinkmann et al. 1971; see first row of Table 8.1) in a German population. This ratio is zero at equilibrium—when selection cannot change the genetic frequencies.

Note that, as commented in previous chapters (Chaps. 2 and 7), the index commonly used to reflect the action of selection is the (narrow sense) heritability, $h^2 = V_A/V_P = V_A/(V_G + V_E)$. Thus, the heritability itself would be below the ratio

Table 8.1 Observed frequencies used in the analyses and results obtained for the different genotypes—fitness values minimizing V_A/V_G and equilibrium frequencies of these fitness values (values taken from Álvarez-Castro and Yang 2011)

	ACP1 genotypes					
	AA	AB	BB	AC	BC	CC
Observed frequencies[a], p_{ij}	0.1242	0.4139	0.3349	0.0445	0.0799	0.0025
Fitness values, ω_{ij}	0.9303	1	0.9667	1.0465	0.9658	0.5832
Equilibrium frequencies, p^*_{ij}	0.1115	0.4323	0.3766	0.0295	0.0491	0.0001

[a] Brinkmann et al. (1971)

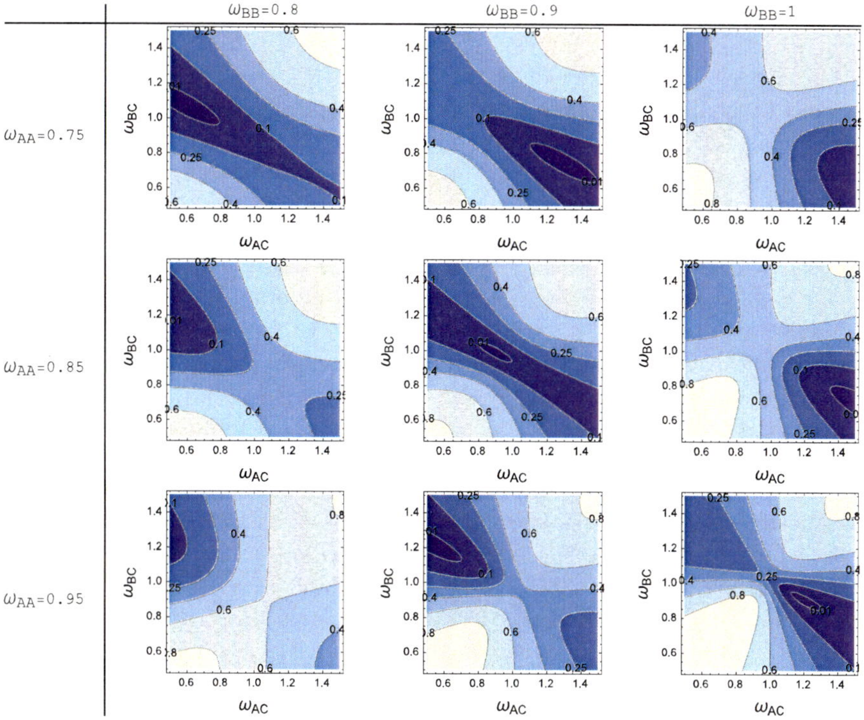

Fig. 8.1 From Álvarez-Castro and Yang (2011). Contour plots of the ratio V_A/V_G for various fitness values in accordance with the verbal model by Greene et al. (2000), with the frequencies observed by Brinkmann et al. (1971). For all panels, $\omega_{AB} = 1$ and $\omega_{CC} = 0.6$

analyzed. In any case, we are looking for polymorphic equilibria where, ideally, $V_A/V_G = h^2 = V_A = 0$. In practice, none of the three variables is expected to attain an exact value of zero, due to the unavoidable imprecision made in obtaining genotype frequencies in the field. Finally, in what regards the concerns raised about the concept of heritability in the aforementioned previous chapters, note that no additional additive variances (V_{AA}, V_{AAA}…) can possibly interfere this index in a one-locus model.

Figure 8.1 (from Álvarez-Castro and Yang 2011) shows that combinations of fitness values in accordance with Greene et al. (2000) verbal model lead to very low V_A/V_G ratios. For instance, the central panel shows that $V_A/V_G < 0.01$ in a region including fitness values of $\omega_{AA} = 0.85$, $\omega_{AB} = 1$, $\omega_{BB} = 0.9$, $\omega_{AC} = 0.9$, $\omega_{BC} = 1$, and $w_{CC} = 0.6$, and a larger region with such low ratio is shown in the panel above it. This way, it gets proved numerically that Greene et al. (2000) verbal model may actually lead to the maintenance of the *ACP1* three-allele polymorphism in Europe and, therefore, that they may indeed have identified the causes of such polymorphism.

8.2.2 Fitness Estimates from Equilibrium Frequencies

In the above, the versatility of NOIA enabled a decomposition of the genetic variance in a multiallelic case and in the face of departures from the Hardy-Weinberg proportions. Accounting for these departures for studying selection has been crucial. Indeed, they are expected as an outcome of selection and forgoing them would be equivalent to disregarding a consequential gear of an engine. More specifically, if the very low V_A/V_G ratios were found using a theory not accounting for departures from the Hardy-Weinberg proportions, the doubt would remain as to whether that result would hold when accounting for the departures. In what follows, having access to Álvarez-Castro and Yang's (2011) decomposition of the genetic variance with multiple alleles and departures from the Hardy-Weinberg proportions proves even more providential.

Ultimately, each variance component is given by a function whose domain is a set of fitness values (or phenotypes, as long as they can be interpreted as fitness values, i.e., assuming directional selection) and a set of genotype frequencies. Such function can thus be minimized for a set of frequencies, the result being the set of fitness values that provide the lowest additive variance possible for that set of frequencies. Note that precision is key in this process, since slightly different sets of frequencies will lead to very different fitness vales—hence the need to account for departures from the Hardy-Weinberg proportions that are caused by the selection processes itself.

Álvarez-Castro and Yang (2011) followed this approach by setting (without loss of generality) a reference fitness of $\omega_{AB} = 1$ and applying two constraints in accordance to Greene et al. (2000) verbal model. First, the genotypes not affected by fetal macrosomia must keep on being sorted as their enzyme activities, which translates into $\omega_{CC} < \omega_{BC} < \omega_{BB}, \omega_{AC}$. Second, the genotypes with extreme enzyme activities (particularly the lowest one) must perform worse than the reference fitness, but not extremely bad, which translates into $0.5 < \omega_{CC} < \omega_{AA} < \omega_{AB} = 1$. The resulting set of frequencies, pursued as the best candidate to maintain the polymorphism at the observed frequencies, is shown in Table 8.1 (from data in Álvarez-Castro and Yang 2011) and in Fig. 8.2 (see also a previously published similar figure by Álvarez-Castro 2016).

Álvarez-Castro and Yang (2011) performed several additional analyses in order to evaluate how good a candidate for explaining the observed frequencies the set of estimated fitness values is. Amongst them, it is first of all noteworthy that the additive variance resulting from that set of frequencies and the observed frequencies actually was obtained as virtually zero. This is indeed reassuring about the potential of the estimated set of fitness values to lie close to neighborhood of an equilibrium point (from which selection does not cause significant genotype frequency changes).

And more to the point, Álvarez-Castro and Yang (2011) simulated the evolution of the system from the set of observed frequencies, using recursions fed with the estimated set of fitness values, assuming random mating, no drift, and non-overlapping generations (see, e.g., Li 1976). The result was a new set of frequencies (see again Table 8.1) from which the system did not evolve further—

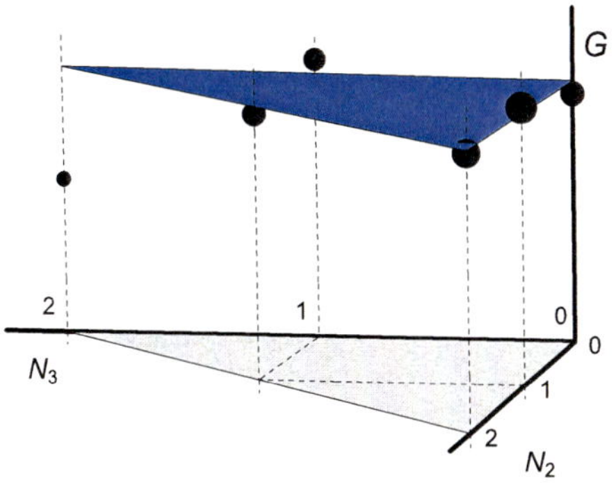

Fig. 8.2 Adapted from Álvarez-Castro (2016). One-locus three-allele GP (strictly speaking, genotype-to-fitness) map obtained by minimizing the additive variance under certain conditions, as explained in the text. The horizontal axes N_2 and N_3 represent the contents of alleles A_2 and A_3, respectively. The blue plane is the weighted (by the genotypic frequencies) linear approximation to the genotypic values (dark circles), and it is parallel to the grey horizontal plane of the coordinates of the genotypes, reflecting a virtually nil additive variance (for further details see the text and Álvarez-Castro and Yang 2011; Álvarez-Castro and Le Rouzic 2015)

an actual stable polymorphic equilibrium point. This equilibrium point does not depart much from the (initial) observed genotype frequencies. Thus, the differences can easily be explained by the consubstantial lack of precision of the sampling process, which affects not only the frequencies themselves but the fitness values from them estimated as well.

To summarize the results so far, the fitness estimates obtained from the observed genotype frequencies satisfactorily explain the maintenance of the three-allele equilibrium of the *ACP1* enzyme in Europe. The fact that it is not expected that any random set of three-allele polymorphic genotype frequencies is very close to equilibrium frequencies (e.g., Lewontin et al. 1978) adds up to the interpretation that the observed frequencies actually reflect a polymorphic equilibrium and that, therefore, the fitness estimates obtained by Álvarez-Castro and Yang (2011) are reliable. Furthermore, since the theoretical approach used is mathematically exact (in particular, it accounts for departures from the Hardy-Weinberg proportions), the only inaccuracies should be due to those of the empirical data.

Nevertheless, Álvarez-Castro and Le Rouzic (2015) have noted that current medical advances (particularly, modern pregnancy monitoring and high expertise in caesarian sections) could be weakening one of the two opposed evolutionary forces identified by Greene et al. (2000), hence unleashing the rise of some of the genotypes—i.e., modifying the fitness values and, thus, the evolutionary properties of the system. Such an evolutionary change could lead to a rapid fixation of the *ACP1*A* allele. Figure 8.3 (by Álvarez-Castro and Le Rouzic 2015) shows indeed how small increases in fitness of the *ACP1*A/ACP1*A* genotype (ω_{AA}) result in a rapid burst of additive variance, which would lead to significant changes of the genotypic frequencies along generations. Thus, in the end, it looks sensible to

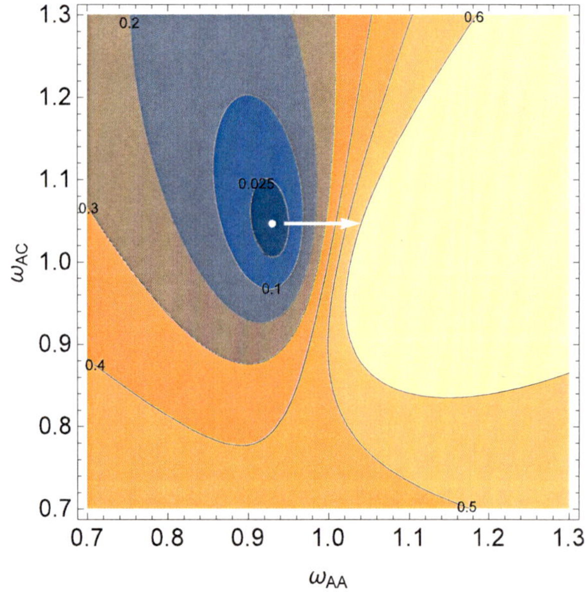

Fig. 8.3 From Álvarez-Castro and Le Rouzic (2015). Contour-plot of the ratio V_A/V_G for varying values of ω_{AA} and ω_{AC}, the remaining fitness values being those in Table 8.1. The white spot marks the minimum value (virtually zero), and the arrow indicates the direction of increasing values of ω_{AA}, along which the additive variance rapidly boosts

assume that the maintenance of the polymorphism is vulnerable due to changes in fitness values caused by recent environmental changes.

8.3 Bateson-Dobzhansky-Müller Incompatibilities

We shall now consider BDM incompatibilities, which is a case of historical evolutionary interest (see e.g. Dobzhansky 1937). With this case, we aim to illustrate the use of orthogonal decomposition of the genetic variance with epistasis and departures from linkage equilibrium. In brief, BDM incompatibilities arise when a population is split into two reproductively isolated populations, which are then invaded by mutations in different loci so that their coexistence in individuals proves detrimental after a secondary contact. The incompatibility of the two new alleles implies epistasis and the secondary contact implies strong departures from linkage equilibrium.

As expounded in Chap. 6, the developments enabling a decomposition of the genetic variance with departures from linkage equilibrium frequencies stem from the work of Álvarez-Castro and Crujeiras (2019). In this work, the theory provided is illustrated by showing that departures from both the Hardy-Weinberg proportions and linkage equilibrium contribute to the emergence of a plateau of additive variance that may prevent the fixation of the original genotype after the secondary contact, thus buying time for the development of mechanisms of sympatric reproductive isolation. We hereafter reproduce that analysis (with just a few modifications).

BDM incompatibilities can be modeled using the simplest possible epistatic model of two biallelic loci. We consider in particular a population in which alleles

Table 8.2 Modified from (Álvarez-Castro and Crujeiras 2019). Genotypic values (in trait units) of the BDM case considered in the text and individual-referenced genetic effects (also in trait units) from which it can be built. It is a locus-symmetric, case, meaning that the marginal (additive and dominance) individual-referenced effects of loci A and B are equal (i.e., $a_1 = a_2 = a$ and $d_1 = d_2 = d$), as well as the individual-referenced pairwise interaction terms da and ad. The reference point, R, is chosen in a way that enables to express the GP map with nil marginal effects. For details about the models used to translate between GP maps and individual-referenced (also called functional) genetic effects, see Chap. 4

Genotypic values				Genetic effects at the individual level		
	A_1A_1	A_1A_2	A_2A_2	R	a	d
B_1B_1	4	4	4	G_{1111}	0	0
B_1B_2	4	0	0	aa	ad	dd
B_2B_2	4	0	0	−1	−1	1

A_1 and B_1 are fixed, which splits into two isolated populations that are in their turn invaded by initially neutral mutations A_2 and B_2, respectively. However, as soon as the two populations enter into secondary contact, the simultaneous occurrence of alleles A_2 and B_2 in individuals causes a fitness decline. Table 8.2 provides the variables of the BDM case we consider here. At the bottom of the table, it is shown that when expressing the BDM GP map in terms of individual-referenced genetic effects from the reference of the genotypic value of $A_1A_1B_1B_1$—$R = G_{1111}$—all marginal effects are nil. Nevertheless, it is well known that the presence of all kinds of epistatsis components (AA, DA, AD, and DD) implies that non-nil marginal effects shall arise both when representing the GP map from different individual-reference points and when analyzing it at the population level. Hence, it is expected that additive, dominance, and epistasis variance components are non-nil under many conditions.

8.3.1 Decomposition of the Genetic Variance Accounting for Departures from Linkage Equilibrium

Such variance components can be observed in Fig. 8.4a, where the genetic variance decomposition is shown for two sets of allele frequencies under their whole range of possible incidence of linkage disequilibrium. Figure 8.4a considers in particular two cases fulfilling $f(A_1) = f(B_2)$, which is expected to occur at the beginning of the secondary contact described above. In one case (red lines), the individuals are assumed to come in equal numbers from the two populations and therefore $f(A_1) = f(B_2) = ½$, whereas in the other case (blue lines), the number of individuals coming from one of the populations doubles that of the other one and therefore $f(A_1) = f(B_2) = ⅓$.

Álvarez-Castro and Le Rouzic (2015) have observed that under linkage equilibrium, despite the evolutionary importance of epistasis in BDM incompatibilities (potentially leading to speciation), the epistasis variance at secondary contact does not exceed half of the additive variance. With the present analysis, this result gets

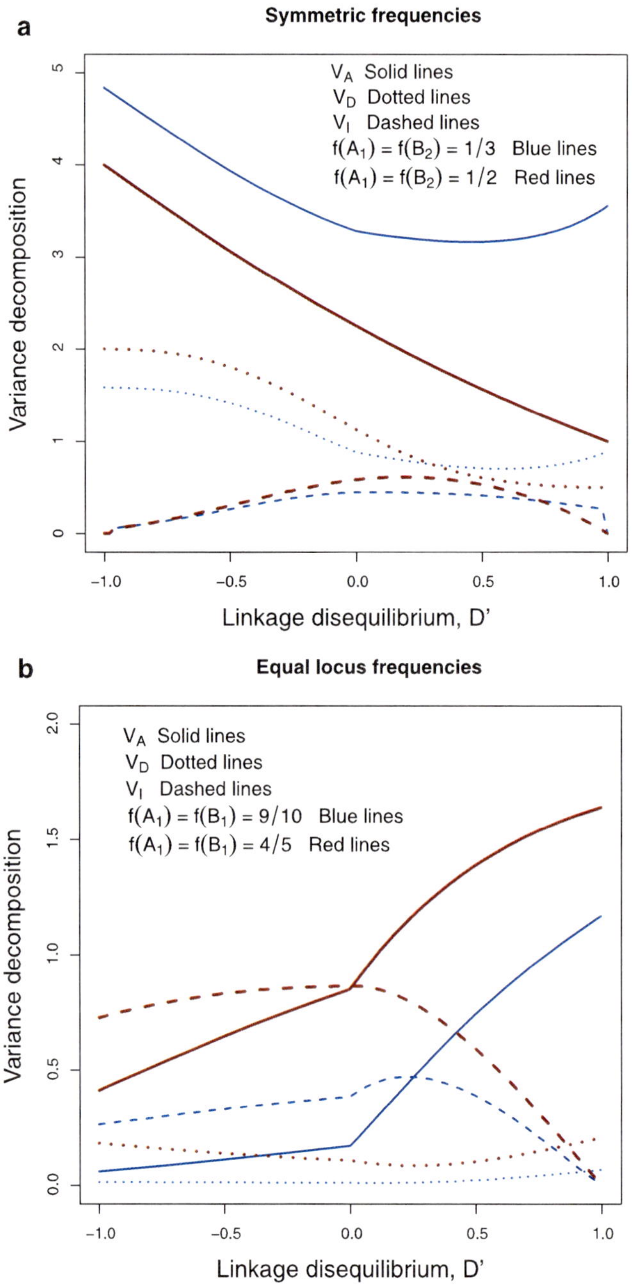

Fig. 8.4 From Álvarez-Castro and Crujeiras (2019). Genetic variance components of the BDM case considered in the text and Table 8.2, for different sets of allele frequencies and for the full range of possible incidences of linkage disequilibrium, measured in terms of the standardized disequilibrium index, D'. Additive, dominance and epistasis variances (in squared trait units) are plotted using solid, dotted and dashed lines, respectively. Panel (**a**) considers two cases with symmetric allele

extended to linkage disequilibrium since, as we can see in Fig. 8.4a, the epistasis variances (dashed lines) remain at values below half of their corresponding additive variances (solid lines) not only when $D' = 0$ (i.e., under linkage equilibrium) but also to the right of that point (i.e., under positive linkage disequilibrium) and to the left of it (i.e., under negative linkage disequilibrium).

The latest case, with negative linkage disequilibrium, is the realistic one after secondary contact, when any individual is expected to produce one only type of gametes, either A_1B_2 or A_2B_1, depending upon its population of origin. Starting with linkage equilibrium (with $D' = 0$, at the centre of Fig. 8.4a) to increasing negative linkage disequilibrium (i.e., towards the left-hand side of the figure), the additive and the dominance variances (solid and dotted lines, respectively) increase. The epistasis variance, on the other hand, decreases. Indeed, linkage disequilibrium makes some multilocus genotypic classes to be underrepresented or even absent in the extreme case, which causes the decrease of the epistatic variance but does not work in the same way for the dominance variance because homozygotes as well as heterozygotes of some kind remain even under maximum negative linkage disequilibrium—e.g., $A_1B_2|A_1B_2$, $A_1B_2|A_2B_1$, and $A_2B_1|A_2B_1$.

The scenario close to fixation of genotype $A_1A_1B_1B_1$ has also been inspected and it is represented in Fig. 8.4b. That figure shows one case with allele frequencies $f(A_1) = f(B_1) = 4/5$ (red lines) and another one with $f(A_1) = f(B_1) = 9/10$ (blue lines). As opposed to Fig. 8.4a, in Fig. 8.4b, the additive variances decrease with increasing incidence of negative linkage disequilibrium. This reveals the extent to which negative linkage disequilibrium (which is expected to remain for a number of generations after secondary contact, particularly if the individuals of the two populations do not freely intermingle) is hindering fixation, thus bestowing extra time for speciation to be triggered—e.g., for additional reproductive isolation (as mating preference mechanisms) to evolve.

That slow down of the selection speed towards fixation due to departures from equilibrium frequencies can also be visualized in Fig. 8.5, which shows several additive variance surfaces of the BDM case here considered. In Fig. 8.5a, equilibrium frequencies are assumed. Figure 8.5b shows a case of negative linkage disequilibrium, with a standardized disequilibrium index of $D' = -0.6$. A decrease in additive variance around the fixation of $A_1A_1B_1B_1$ can be perceived as an incipient plateau towards the left corner of the additive variance surface in Fig. 8.4b, as compared with Fig. 8.5a. That plateau would become much more evident with increasing incidences of negative linkage disequilibrium, as Fig. 8.4b demonstrates.

In Fig. 8.5c, a different kind of departures from equilibrium frequencies is shown—departures from Hardy-Weinberg. Indeed, also a reduction of heterozygotes is expected in the BDM case at secondary contact and as well in the

Fig. 8.4 (continued) frequencies across loci, $f(A_1) = f(B_2) = 1/3$ (blue lines) and $f(A_1) = f(B_2) = 1/2$ (red lines). Panel (**b**) considers two cases with equal allele frequencies across loci, $f(A_1) = f(B_2) = 9/10$ (blue lines) and $f(A_1) = f(B_2) = 4/5$ (red lines)

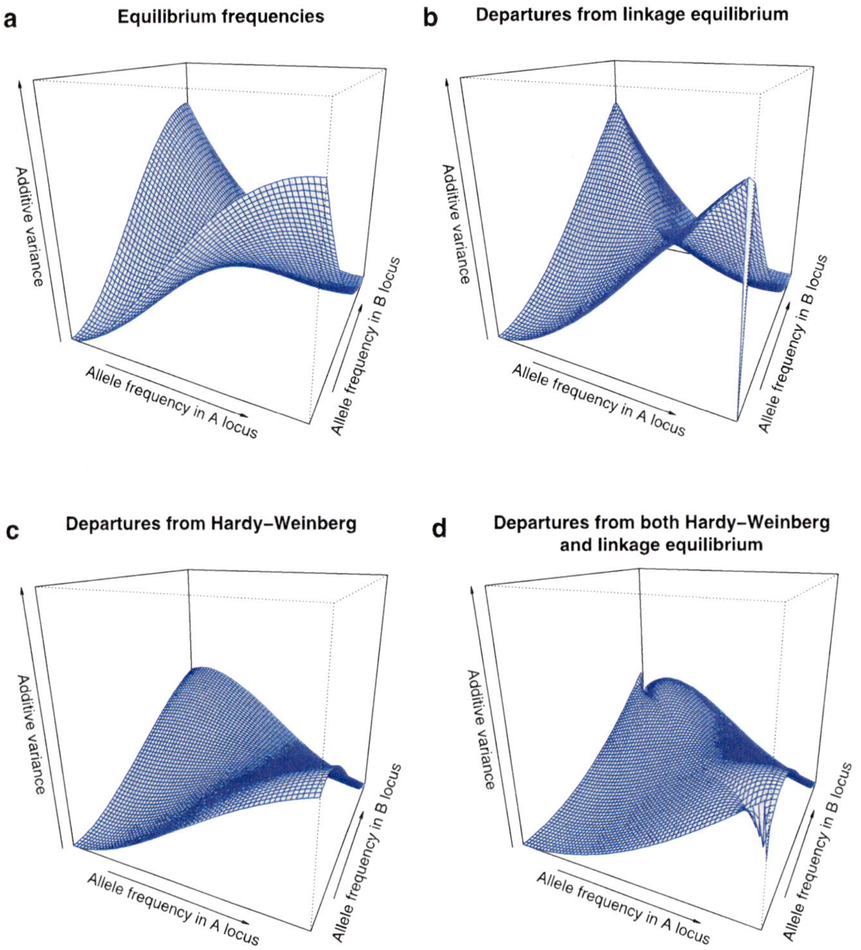

Fig. 8.5 From Álvarez-Castro and Crujeiras (2019). Additive variance (in squared trait units) surface of the BDM case considered in the text and Table 8.1, for the whole range of allele frequencies and different incidences of departures from equilibrium frequencies. Panel (**a**) considers the case of equilibrium frequencies. Panel (**b**) considers the case of negative linkage disequilibrium with $D' = -0.6$. Panel (**c**) considers the case of a reduction of heterozygotes with a fixation index of $F = 0.3$ at each locus. Panel (**d**) combines both departures (of Panel (**b**) and Panel (**c**)) at the same time. In all cases the vertical axis ranges from zero to ten

following generations as long as the two populations do not freely intermingle. We have assumed in particular a fixation index of $F = 0.3$ at each of the two loci. That incidence is enough to cause an evident additive variance plateau around the fixation of $A_1A_1B_1B_1$ (in particular, more evident than in the case of $D' = -0.6$ in Fig. 8.4b). In Fig. 8.4d, with the combined effect of linkage disequilibrium (as in Fig. 8.4b) and departures from Hardy-Weinberg (as in Fig. 8.4c), the additive variance plateau becomes even larger. Therefore, Fig. 8.5 shows how non-equilibrium frequencies may hamper fixation to occur in a case of BDM incompatibilities, in terms of the changes those departures from equilibrium frequencies (both from Hardy-Weinberg and from linkage equilibrium) cause in the additive variance.

8.3.2 Assessing the Contribution of the Epistatic Variance to Selection Response

As discussed in Chap. 2, the epistatic components of the genetic variance implying only additive effects (V_{AA}, V_{AAA}...) are also known to contribute to the narrow sense heritability, h^2, even when it is overly expressed just as $h^2 = V_A/V_G$. It is particularly convenient to keep this in mind when computing the decomposition of the genetic variance from the knowledge of the GP map (although it also has some implications when estimating h^2 from empirical data on phenotypes and relatedness). Thus, Álvarez-Castro and Crujeiras' (2019) analysis is here completed with an inspection of V_{AA}—the only additional component to consider in the study of heritability of a two-locus system.

Figure 8.6 shows V_{AA} in the same way as Fig. 8.5 showed V_A, with the exception of a 10-fold difference in scale. This has been done for facilitating the visualization, since the values V_{AA} attains in Fig. 8.6 (with the vertical axis ranging from zero to one) are much lower than those of V_A in Fig. 8.5 (with the vertical axis ranging from zero to ten). From Expression 2.38, in particular with $r = ½$ and $u = 0$, the effect of V_{AA} in the heritability can be found to be half of that of V_A. Thus, altogether, the effect of V_{AA} in the heritability is twenty times lower than what could be inferred by comparing Figs. 8.4 and 8.5 at first sight.

Thus, Fisher's (1918) assumption of a minor effect of the epistatic components of variance (discussed in Chap. 2) holds for this case. There are two very important remarks to keep in mind in relation with this. First, it is important not to forget that the interaction variances can be low even in cases where interactions are playing an important evolutionary role. In Fig. 2.6, we have seen that for one biallelic locus, overdominance (which conditions the maintenance of the polymorphism) leads to the dominance variance to be lower than the additive variance for over 40% of the possible allele frequencies. In the present case of BDM incompatibilities, the epistatic variance components are much lower than the additive variance despite epistasis being the evolutionary mechanism potentially leading to not less than speciation. Incidentally, this result was already pointed out with empirical support (Álvarez-Castro et al. 2012).

Second, neither one nor several particular cases build up a general theory. In relation with this, it is that interaction variances are expected to decrease towards the fixation points, with decreasing genetic variation and hence smaller effects of the interactions on average over the population. This is illustrated, for instance, by the aforementioned Fig. 2.6 of Chap. 2. But there is no sound reason to support that the interaction variances may not be high otherwise. Indeed, cases in which the epistatic variance components entail a significant portion of the genetic variance have been reported. In point of fact, such cases may be of high evolutionary importance.

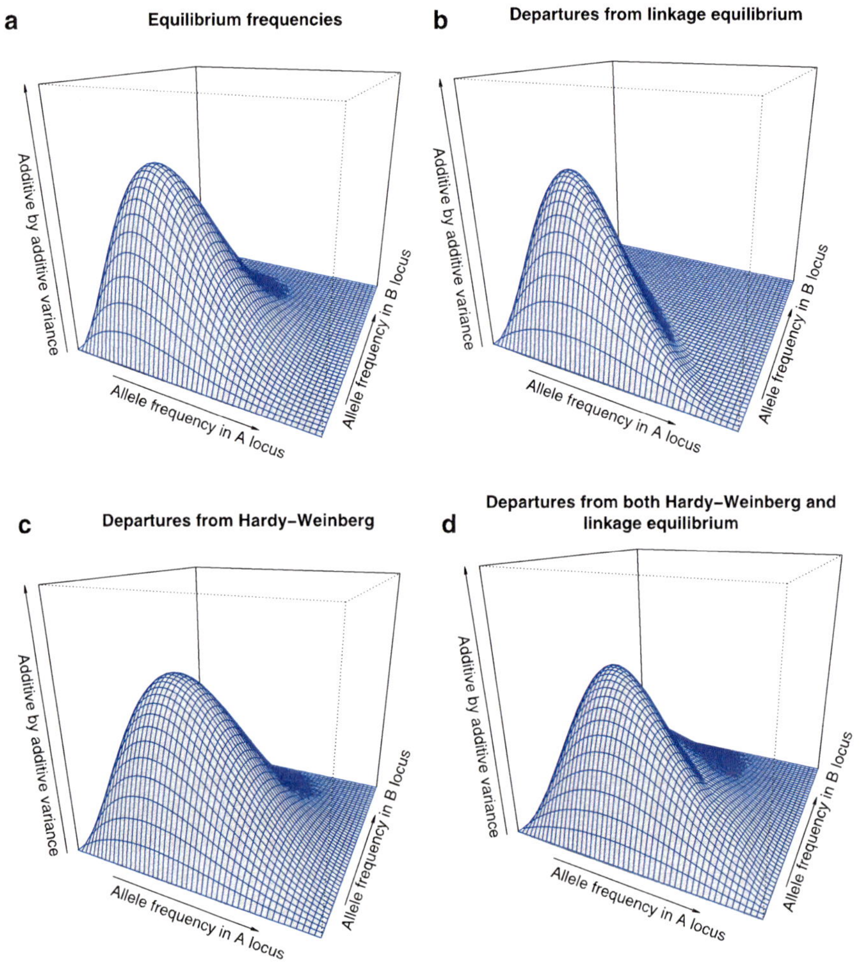

Fig. 8.6 Additive-by-additive variance (in squared trait units) surface of the BDM case considered in the text and Table 8.1, for the whole range of allele frequencies and different incidences of departures from equilibrium frequencies. All panels follow the same pattern as in Fig. 8.5, but with a different scale at the vertical axis, which here ranges from zero to one (instead of from zero to ten)

Epistasis generates rugged evolutionary landscapes, with many polymorphic equilibria including local maxima, local minima, and saddle points. The neighborhoods of the latest ones are evolutionary plateaus that slow down evolution rates, with high proportions of the genetic variance accounted for by epistasis components (Álvarez-Castro and Le Rouzic 2015; Goodnight 2015; Le Rouzic and Álvarez-Castro 2016).

In any case, Fig. 8.6 shows that departures from equilibrium frequencies reduce the values of the function around the corner at the left of the panels, similar to what has been explained in detail above in relation to Fig. 8.5. As we have made it clear above that the epistatic variance is for this case very low (despite being exaggerated visually in Fig. 8.6 just to perceive its shape), this reduction in additive-by-additive variance does not have a significant effect compared with that of the additive variance. To end up with, it is worthwhile recalling that all computations made for this analysis can be adequately performed only as long as advanced models of genetic effects properly implementing general genetic architectures and arbitrary departures from equilibrium frequencies are available (and used), which is the main objective of this book.

8.4 Gene-Environment Interaction in Precision Medicine

Gene-environment interaction is key to the understanding of a large spectrum of human disorders ranging from obesity, cardio-metabolic diseases, and other metabolic disorders to cancer, autoimmune diseases, and mental disorders (e.g. Karl and Arnold 2015; Lopizzo et al. 2015; Flouris et al. 2017; Cust 2020; Smith 2020; Teperino 2020). In this regard, medical advances have been hindered due to the lack of theoretical developments appropriately accounting for the interplay between gene-environment interaction and gene-environment correlation (Assary et al. 2020, see also Chap. 6).

It is indeed to be expected that gene-environment interaction occurs together with non-random associations between genes and environments both in what regards human health (since individuals with genetic risk under certain environments will particularly wish to avoid them) and general evolutionary biology (since individuals will try to stick to the environments where they perform best). Opportunely, the theory that was required for filling the aforementioned gap has already been developed (Álvarez-Castro 2020) and presented in Chap. 6. In this section, we illustrate some potential use of that theory, particularly in the field of medicine (modified from Álvarez-Castro 2020).

Advances in genetic methodologies have latterly been applied to increase accuracy of medical predictions, particularly in what regards treatment efficiencies and prevention strategies for different (groups of) individuals. This approach has been coined as precision medicine. Within this framework, we shall analyze two built-in cases on disease susceptibility with environmental influence.

8.4.1 Cases of Disease Susceptibility Under Environmental Exposure

The first case we here consider is a case of genetic risk to disease of one biallelic locus under two different environmental conditions, built up by Li et al. (2019). We shall refer to this case as the risk and environment (RAE) case (see Table 8.3). Applying the formulation of NOIA at the individual level (see Chap. 4) from the default (non-exposed and non-risk) individual reference (0.01), the additive, dominance, environment, additive-by-environment, and dominance-by-environment

Table 8.3 From Álvarez-Castro (2020). Phenotypes (disease susceptibility) of the four individual classes (risk allele carriers and non-carriers under exposed and non-exposed environments), for the two cases considered in the text—the case taken from Li et al. (2019), here called the risk and exposure (RAE) case and the genetic risk to exposure (RTE) case. Complete dominance of the risk allele is assumed so that homozygotes for the risk allele and heterozygotes are equally susceptible to the disease

		Genetics	
Case	Environment	Default	Risk
RAE	Default	0.01	0.5
	Exposed	0.4	0.9
RTE	Default	0.01	0.01
	Exposed	0.4	0.9

effects reflecting the aforementioned substitutions are 0.245, 0.245, 0.39, 0.005, and 0.005, respectively. Those values show that the RAE case entails both genetic and environmental effects with extremely small gene-environment interaction effects, relative to both the genetic and the environmental marginal contributions. More specifically, the interaction effects actually lay about two orders of magnitude below the marginal effects. Thus, the case can hardly be considered a gene-environment interaction case as originally intended.

Let us now move on to the second case, which can be described as a case of (genetic) risk to (environmental) exposure (thus referred to hereafter as RTE). In particular, the risk allele increases disease susceptibility only when combined with exposure, hence actually interacting with the environment. Thus described, the interaction behaves as a switch—the environmental effect shall either be switched on (when carrying the risk allele) or turned off (otherwise), as shown in Table 8.3.

Thus defined, the RTE involves notable gene-environment interaction, as opposed to the previous case—the RAE. Specifically, the additive and dominance effects (i.e., the marginal genetic effects) at the individual level of the RTE case from the reference of the individual default class (no genetic risk and no exposure) are zero, which actually is in accordance with the genetic risk being turned off in the absence of exposure. The environmental interaction effects, on the other hand, are notable—in particular, 0.25 both in what regards the additive- and the dominance-by-interaction effects.

8.4.2 Predictions Under Varying Exposure

Li et al. (2019) defined the "genetic coefficient of the disease" (at a population) as the index accounting for the "difference between the additive expectations of case genomes and control genomes." In the models we are working with, the index reflecting that biological meaning is 2α, which reflects that it is measured in units of disease susceptibility (since the genetic effects are measured in trait units). Figure 8.7 shows the genetic coefficient of the disease (i.e., 2α), computed using

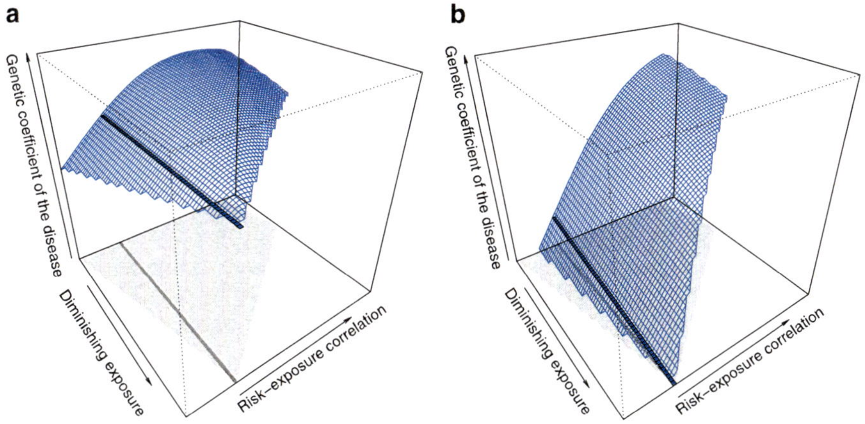

Fig. 8.7 Modified from Álvarez-Castro (2020). Genetic coefficient of the disease obtained with ARNOIA. Panel (**a**) corresponds to the RAE case and panel (**b**) does to the RTE case. The risk allele frequency is 0.15, with genotypic frequencies under Hardy-Weinberg equilibrium. The environmental exposure frequency ranges from 0.24 to zero. For each environmental exposure frequency, the whole range of possible correlations (including both positive and negative associations) of the risk allele and environmental exposure are considered. The range of possible risk-exposure correlations is shown by the light gray area at the bottom of the figure (in the plane where the coefficient of the disease is nil). The values of the vertical axis range from zero to 1.20, although in Panel (**b**) the genetic coefficient of the disease takes negative values for strong negative correlations (towards the left of the figure) that are not shown (as they are below the light grey area). The thick blue line (whose projection at the bottom of the figure is shown in dark gray) marks the absence of correlation between risk allele and environmental exposure, which are the ones the original NOIA setting (not accounting for non-random associations between genes and environment) would provide for the whole range of correlations between the risk allele and environmental exposure

ARNOIA (as derived in Chap. 6), for the RAE and the RTE cases (panels A and B, respectively) under a hypothetical decrease of the environmental exposure and for the different possible ranges of risk-exposure correlation.

The thick blue line in Fig. 8.7a marks random association and shows that the genetic coefficient of the disease is simply not affected by decreasing the exposure frequency in the population. This is as expected under lack of interplay between gene an environment (i.e., no interaction and no correlation). Indeed, although the trait is subject to both genetic and environmental influence, as long as there is no (or very little) interplay between them, the genetic parameter remains virtually constant in the face of variations in the environmental exposure.

Figure 8.7a also illustrates that the coefficient of the disease is notably affected by risk-exposure correlations even in the absence of significant gene-environment interaction. In particular, negative associations between the risk allele and the degree of exposure make the genetic coefficient of the disease to decrease, as the surface to the left of the thick blue line shows. Conversely, positive associations make it to increase, to the right of the thick blue line, although this occurs up to a maximum followed by a slight decrease. Note also that the range of possible risk-exposure correlations (shown by the light gray area at the bottom of the figure) narrows down as the exposure frequency approaches zero, which explains the tip of the surface at the end of the thick blue line.

In Fig. 8.7b, the RTE case is displayed in a way analogous to the RAE case in Fig. 8.7a. As Fig. 8.7b shows, for the RTE case, the genetic coefficient of the disease decreases for decreasing values of exposure under random association of risk and environment (decreasing thick blue line). This coefficient also decreases for decreasing (increasingly negative) association between the risk allele and environmental exposure, as the left tip of the surface shows. In plain language, the figure shows that the problem of increased disease susceptibility of the carriers of the risk allele may be reduced either by reducing exposure for the whole population or by restricting the access to the exposed environment only for the risk population or even though any intermediate alternative (any reduction of the exposure in the population biased towards the carriers of risk alleles).

Note indeed that beyond decreasing, the genetic coefficient of the disease cancels out and even takes negative values with increasingly negative risk exposure correlation (these negative values of the blue surface cannot be seen in Fig. 8.7b because they drop below the grey surface). This makes perfect sense in the RTE case as the risk allele at the default environment outperforms the default allele at the exposed environment—which is not the case in the RAE case (see Table 8.3). In any case, the optimal management of the RTE case depends just upon the reluctance of the average individual to avoid the exposed environment (or even the actual feasibility of bringing the whole population out of it) and the cost of tests to detect the risk allele, which enable personalized warnings.

Overall, the RAE and RTE cases considered in Fig. 8.7 deal with rather singular instances (virtually absent and switch-type, respectively) of gene-environment interaction, for which verbal predictions would be feasible even without mathematical modeling. The results obtained using ARNOIA not only reassuringly agree with the conceptually attainable predictions but also further illustrate how to precisely quantify any desired genetic/environmental parameter. Such advantage can hereinafter be applied to more complex real cases of interest, including more complex cases undergoing less intuitive behaviors.

8.5 Advances in Evolutionary Quantitative Genetics

The cases considered in this chapter illustrate the practical use of some of the most important theoretical advances provided in the previous chapters of this book. The three cases are in fact different in many ways. The first case uses real data to analyze the maintenance of a polymorphism, the second one deals with a historical proposal to explain speciation, and the third one analyzes built-in situations that exemplify those that are currently subject of cutting-edge medical research. They also involve different genetic architectures—multiple alleles, epistasis, and gene-environment interactions, respectively. Nevertheless, the three cases have in common that departures from the equilibrium frequencies are key to an insightful analysis of each of them.

In the first case, it is necessary to account for departures from Hardy-Weinberg to study the action of selection (and, particularly, to aim reasonable estimates of fitness values), since those departures are precisely the evidence to follow. In the second case, beyond the aforementioned departures, a proper account of departures from linkage equilibrium is also crucial in order to convincingly analyze their role in the speciation processes. And finally, departures from random associations between genes and environments are to be accurately taken into account in the third case in order not to confound them with the gene-environment interaction effects to be studied. It is only by disentangling both mechanisms that estimates enabling precise strategies to confront a disease with both genetic and environmental causes can be attained.

The three cases also share the fact that they use a methodology developed within the framework of quantitative genetics to address issues falling within the conventional boundaries of population and evolutionary genetics. Indeed, both the maintenance of polymorphic equilibria under selection, fitness estimation, speciation, and analyzing the genetic component of human health affections at the population level have been addressed from the population and evolutionary genetic perspective. Thus, the applied cases here presented contribute to the merge of both fields, referred to as evolutionary quantitative genetics. We are hereby stressing that the cases here considered show that the implementation of departures from equilibrium frequencies in quantitative genetic methods determine whether they are applicable to aid population and evolutionary genetics.

8.5.1 Implementing Departures from Equilibrium Frequencies Fuel Evolutionary Quantitative Genetics

In the first case of the *ACP1* human enzyme, interactions were restricted to just dominance. Multiple alleles increases the complexity moderately, with the kth allele adding up k new variables to the already existing ones with $k - 1$ alleles. Indeed, considering only three alleles already makes the equilibrium properties of the system to be very complex—whereas in the biallelic case they were easily reduced to a simple expression. In the approach presented above, the quantitative genetics tool of the variance decomposition has been used in sort of a reverse manner. Defining the variance decomposition (a quantitative genetics tool) as an accurate function of the genotypic values and the genotypic frequencies enabled to obtain fitness estimates (an evolutionary genetics index) as the genotypic values minimizing the observed genotypic frequencies. NOIA (particularly, ARNOIA) provides such a function under much more complex genetic architectures than one locus with three alleles, which makes this procedure widely applicable.

Addressing BDM incompatibilities requires to consider (at least) two loci with epistatic interactions. With epistasis, the nth biallelic locus adds up $3^n - 3^{n-1}$ new variables to the already existing ones—i.e., it multiplies times three the previous number of variables. Moreover, the properties of the genetic system get much more cumbersome when the genotypic frequencies depart from linkage equilibrium, which (as opposed to the previous case) makes it necessary to apply ARNOIA (derived in Chap. 7) for obtaining an accurate decomposition of the genetic variance. Indeed, Álvarez-Castro and Crujeiras (2019) pointed out that epistatic systems under departures from linkage equilibrium display emerging properties. In particular, additive-by-additive epistasis alone may generate dominance variance even in the absence of dominance.

The level of complexity coming from the merger of epistasis and departures from linkage equilibrium seems to have made researchers to seriously doubt for a long time that a satisfactory implementation of epistatic systems with departures from linkage equilibrium could be attained at all (see Chap. 7 for further details). The variance decomposition has been already used to study the equilibrium properties of epistatis systems beyond the BDM case, as pointed out above. The accurate decomposition of the genetic variance under departures from linkage equilibrium presented in Chap. 7 makes it now possible to upgrade the accuracy of the use of quantitative genetics tools in such analyses, as the BDM case presented above exemplifies.

The study of gene-environment interaction was considered crucial in classical population genetics and, in general, in evolutionary biology. Indeed, the reaction norm (the genotypic values of one genotype as a function of the possible environments) was acknowledged as the actual target unit of selection (see, e.g., Sarkar 2004). Under gene-environment interaction, correlation between genes and environmental conditions is expected to occur in many instances, whether in human or natural populations, since individuals may be able to choose the environments in which they perform best. Thus also in this case, advanced modeling of genetic effects makes a great difference in applicability of quantitative genetics tools to population and evolutionary genetic studies, as exemplified above with the illustrative example of the use of ARNOIA in a population analysis in the context of precision medicine.

8.5.2 Improving Previous Analyses and Enabling New Ones

The Lande equation (Lande 1979, 1980; Lande and Arnold 1983)—an extension of the breeders equation to multiple putatively correlated traits—has been claimed to be the "focal point" of evolutionary quantitative genetics (Hansen 2023). In relation with this, it is however noteworthy that the first publications of this discipline (Lande 1976a, b;see Hansen 2023) dealt with only one trait. These papers studied drift, selection and mutation by "placing increased emphasis on phenotypic parameters" (Lande 1976b). In the present days when gene mapping is a common practice, it is possible to analyze the evolution of quantitative traits also with an emphasis on the genetic parameters. But for carrying out that task in an appropriate manner, it is necessary to rely on mathematical developments that are upgraded to the present context. In this book, such developments are provided at the univariate (one trait at a time) level, thus laying the groundwork for future work at the multivariate level as well.

Overall, we have in this chapter provided examples reflecting that NOIA and, whenever necessary, its upgraded version, ARNOIA, make a difference in practice in relation with what could be done with previous theory. Using the quantitative genetic developments provided in this book, previous analyses performed with models of genetic effects that did not completely match the situations under study can now be revised. Further, new analyses of the action of selection can be developed which were previously out of reach, as, for instance, the estimation of fitness values from equilibrium frequencies, as the first case presented above illustrates. This procedure is possible because NOIA is mathematically correct beyond equilibrium frequencies, which actually occur as a consequence of the action of selection to be analyzed. In brief, on account of properly implementing departures from the equilibrium frequencies, the theoretical proposal provided in this book lubricates the evolutionary quantitative genetics machinery.

References

Álvarez-Castro JM (2016) Genetic architecture. In: Wolf JB (ed) Encyclopedia of evolutionary biology. Oxford Academic Press, Oxford, pp 127–135

Álvarez-Castro JM (2020) Gene-environment interaction in the era of precision medicine—filling the potholes rather than starting to build a new road. Front Genet 11:921

Álvarez-Castro JM, Crujeiras RM (2019) Orthogonal decomposition of the genetic variance for epistatic traits under linkage disequilibrium-applications to the analysis of Bateson-Dobzhansky-Muller incompatibilities and sign epistasis. Front Genet 10:54

Álvarez-Castro JM, Le Rouzic A (2015) On the partitioning of genetic variance with epistasis. In: Moore JH, Williams SM (eds) Epistasis: methods and protocols. Springer, Humana Press, New York, pp 95–114

Álvarez-Castro JM, Yang R-C (2011) Multiallelic models of genetic effects and variance decomposition in non-equilibrium populations. Genetica 139:1119–1134

Álvarez-Castro JM, Yang RC (eds) (2015) Models and estimation of genetic effects. Frontiers Media, Lausanne

Álvarez-Castro JM, Le Rouzic A, Andersson L, Siegel PB, Carlborg O (2012) Modelling of genetic interactions improves prediction of hybrid patterns—a case study in domestic fowl. Genet Res (Camb) 94:255–266

Assary E, Vincent J, Machlitt-Northen S, Keers R, Pluess M (2020) The role of gene-environment interaction in mental health and susceptibility to the development of plychiatric disorders. In: Teperino R (ed) Beyond our genes. Pathophysiology of gene and environment interaction and epigenetic inheritance. Springer, Cham, pp 117–138

Brinkmann B, Hoppe HH, Hennig W, Koops E (1971) Red cell enzyme polymorphisms in a northern German population. Gene frequencies and population genetics of the acid phosphatase (AP), phosphoglucomutase (PGM), adenylate kinase (AK), adenosine deaminase (ADA) and 6-phosphogluconate dehydrogenase (6-PGD). Hum Hered 21:278–288

Cockerham CC (1954) An extension of the concept of partitioning hereditary variance for analysis of covariances among relatives when epistasis is present. Genetics 39:859–882

Cust AE (2020) Gene-environment interactions and melanoma risk. Br J Dermatol

Dobzhansky T (1937) Genetics and the origin of species. Columbia University Press, New York

Eze LC, Tweedie MC, Bullen MF, Wren PJ, Evans DA (1974) Quantitative genetics of human red cell acid phosphatase. Ann Hum Genet 37:333–340

Fisher RA (1918) The correlation between relatives on the supposition of Mendelian inheritance. Trans Roy Soc Edinburgh 52:339–433

Flouris AD, Shidlovskii YV, Shaposhnikov AV, Yepiskoposyan L, Nadolnik L, Karabon L, Kowalska A, Carrillo AE, Metsios GS, Sakellariou P (2017) Role of UCP1 gene variants in interethnic differences in the development of cardio-metabolic diseases. Front Genet 8:7

Goodnight C (2015) Long-term selection experiments: epistasis and the response to selection. In: Moore JH, Williams SM (eds) Epistasis. Methods and protocols. Springer, New York, pp 1–18

Greene LS, Bottini N, Borgiani P, Gloria-Bottini F (2000) Acid phosphatase locus 1 (ACP1): possible relationship of allelic variation to body size and human population adaptation to thermal stress-a theoretical perspective. Am J Hum Biol 12:688–701

Hansen TF (2023) Variation, inheritance, and evolution: a primer on evolutionary quantitative genetics. In: Hansen TF, Houle D, Pavlicev M, Pélabon C (eds) Evolvability: a unifying concept in evolutionary biology? The MIT Press, Cambridge, pp 73–100

Hopkinson DA, Spencer N, Harris H (1963) Red cell acid phosphatase variants: a new human polymorphism. Nature 199:969–971

Karl T, Arnold JC (eds) (2015) Shizophrenia: a consequence of gene-environment interactions? Frontiers Media SA, Lausanne

Kempthorne O (1954) The correlation between relatives in a random mating population. Proc R Soc Lond B Biol Sci 143:102–113

Kempthorne O (1957) An introduction to genetic statistics. Wiley, New York

Kempthorne O (1968) The correlation between relatives on the supposition of Mendelian inheritance. Am J Hum Genet 20:402–403

Kempthorne O (1997) Heritability: uses and abuses. Genetica 99:109–112

Kimura M (1956) Rules for testing stability of a selective polymorphism. Proc Natl Acad Sci U S A 42:336–340

Lande R (1976a) The mainteneance of genetic variability by mutation in a polygenic character with linked loci. Genatical Research 26:221–235

Lande R (1976b) Natural selection and random genetic drift in phenotypic evolution. Evolution 30: 314–334

Lande R (1979) Quantitative genetic analysis of multivariate evolution, applied to brain:body size allometry. Evolution 33:402–416

Lande R (1980) The genetic covariance between characters maintained by pleiotropic mutations. Genetics 94:203–215

Lande R, Arnold SJ (1983) The measurement of selection on correlated characters. Evolution 37: 1210–1226

Lander ES, Botstein D (1989) Mapping mendelian factors underlying quantitative traits using RFLP linkage maps. Genetics 121:185–199

Le Rouzic A, Álvarez-Castro JM (2016) Epistasis-induced evolutionary plateaus in selection responses. Am Nat 188:E134–E150

Lewontin RC, Ginzburg LR, Tuljapurkar SD (1978) Heterosis as an explanation for large amounts of genic polymorphism. Genetics 88:149–169

Li CC (1967) Genetic equilibrium under selection. Biometrics 23:397–484

Li CC (1976) First course in population genetics. The Boxwood Press, Pacific Grove

Li J, Li X, Zhang S, Snyder M (2019) Gene-environment interaction in the era of precision medicine. Cell 177:38–44

Lopizzo N, Bocchio Chiavetto L, Cattane N, Plazzotta G, Tarazi FI, Pariante CM, Riva MA, Cattaneo A (2015) Gene-environment interaction in major depression: focus on experience-dependent biological systems. Front Psych 6:68

Mandel SPH (1959) The stability of a multiple allelic system. Heredity 12:298–302

Provine WB (1971) The origins of theoretical population genetics. University of Chicago Press, Chicago

Sarkar S (2004) From the reaktionsnorm to the evolution of adaptive plasticity: a historical sketch, 1909-1999. In: Dewitt TJ, Scheiner M (eds) Phenotypic plasticity. Oxford University Press, New York, pp 10–30
Sensabaugh GF, Golden VL (1978) Phenotype dependence in the inhibition of red cell acid phosphatase (ACP) by folates. Am J Hum Genet 30:553–560
Smith M (2020) Gene environment interactions. Nature and nurture in the twenty-first century. Elsevier Academic Press
Spencer N, Hopkinson DA, Harris H (1964) Quantitative differences and gene dosage in the human red cell acid phosphatase polymorphism. Nature 201:299–300
Stoltzfus A, Cable K (2014) Mendelian-mutationism: the forgotten evolutionary synthesis. J Hist Biol 47:501–546
Teperino R (ed) (2020) Beyond our genes. Pathophysiology of gene and environment interaction and epigenetic inheritance. Springer, Cham
Van Der Veen JH (1959) Tests of non-allelic interaction and linkage for quantitative characters in generations derived from two diploid pure lines. Genetica 30:201–232
Wilder JA, Hammer MF (2004) European ACP1*C allele has recessive deleterious effects on early life viability. Hum Biol 76:817–835

9 The Comes and Goes of the Black Box Perspective in Quantitative Genetics

Abstract

Since its birth and for a long time, quantitative genetics assumed the infinitesimal model as a simplistic baseline on which to sustain a top-down approach—it enabled a black box perspective leading to the development of empirical studies in the absence of methods to inspect real genetic architectures. However, the infinitesimal model is in opposition to several relevant evolutionary phenomena—it leads to predictions that have been found to oppose empirical observations. Gene-mapping experiments eventually started to dig into the black box concealing the real genetic architectures, thus entailing a bottom-up approach to assess and complement the previous top-down one. Criticism has been raised about the possibilities of gene mapping and the black box perspective re-emerged in the form of genomic selection. In its turn, the natural and orthogonal interaction (NOIA) model has been used to assess the limitations of genomic selection. It is here shown that advanced genetic modeling provides adequate support for a consensus between the top-down and the bottom-up approaches by feeding the merge of quantitative and population genetics into evolutionary quantitative genetics. Lessons learnt from the early history of genetics are here harnessed for that matter.

9.1 Introduction

Genetics was born right with the twentieth century and generated an intense debate about the role of Mendelian factors in the evolution of traits of continuous variation (Provine 1971). As already expounded in Chap. 2, the work of Fisher (1918) set the cornerstone for the genetic community to comprehend how quantitative traits underlain by Mendelian factors evolve, thus giving birth to quantitative genetics. But it is convenient to keep in mind that Mendelian factors (which we nowadays can refer to as loci) were no more than abstract entities at that time, and they remained so

J. M. Álvarez-Castro, *Genes, Environments and Interactions*,
https://doi.org/10.1007/978-3-031-41159-5_9

for some time to come. In brief, several decades would have to pass for key advances in the identification (Avery et al. 1944; Hershey and Chase 1952) and in the understanding (Franklin and Gosling 1953; Watson and Crick 1953) of their molecular basis of biological heredity to lead to the first decisive steps of DNA sequencing (Wu 1972) and of gene mapping (Lander and Botstein 1989).

Opportunely, Fisher (1918, 1930) identified, in his derivations, indexes reflecting crucial properties of the action of selection that could be obtained from empirical data on phenotypes and relatedness. Such indexes stemmed from the components of the genetic and of the phenotypic variances and made valuable practical applications possible long before the underlying loci could be inspected by any means. This way, the genetic architectures of the traits under study were bypassed in the analyses. When an underlying explanatory hypothesis was needed, it would typically be taken from the infinitesimal model—assuming the action of many loci of small effect and negligible interaction between/amongst loci (see e.g. Bulmer 1980). This is the strategy we have already referred to in Chap. 2 as a black box perspective—predictions about the action of selection could be attained despite of the fact that the effects of the alleles at the loci remained hidden from the eyes of the researcher.

The narrow sense heritability is a paradigmatic example of the aforementioned strategy. As also explained in Chap. 2, the correlation between relatives within one generation has been used to estimate the heritability of the trait at the population under study, from which predictions for the response to selection under different selection intensities have been made. Concerns raised about the reliability of this procedure have been mentioned in Chaps. 2, 7, and 8 and are also resumed in the Addendum of this book. While that procedure may not always provide the most accurate predictions, the point made here is about quantitative genetic methodologies and applications having been developed for a long time in the lack of standard procedures to obtain explicit genetic architectures of the traits under study.

As mentioned above, however, the time eventually came when researchers could start to look inside the black box where those real genetic architectures were long hidden. Analogous to what has been said in the previous paragraph, while the endeavor of reliably dissecting real genetic architectures may not be accomplished in one shot, the point made here is about a paradigm shift changing our perception of the scope of quantitative genetic research. Ultimately, after about two generations of researchers having worked with models of genetic effects, these models needed to be revised for being applied under a completely different perspective—turning our views of genetic architectures from being concealed inside impenetrable black boxes to being wrapped by nuisance veils.

Several of those veils proved tricky and the one the present book is to remove arose from the limitations of the models of genetic effects available at the time (see e.g. Cheverud and Routman 1995; Hansen and Wagner 2001; Yang 2004; Zeng et al. 2005; Álvarez-Castro and Carlborg 2007; Álvarez-Castro and Yang 2015). Indeed, the main objective of this book can be expressed precisely as expounding the theory necessary to upgrade the models of genetic effects for overcoming their limitations in general terms. A number of more specific instances follow.

The models of genetic effects compiled in this book—the natural and orthogonal interactions (NOIA) model—can be used to improve gene-mapping methods, leading to the most plausible genetic architecture underlying the dataset, as discussed in Chap. 7. They can also provide more adequate interpretations of the results found by means of such gene-mapping experiments, particularly taking into account that the quantitative genetic parameters change under different population contexts and that researchers will generally be interested in analyzing populations different from that used to disclose the genetic architecture. The computation of the coefficient of a disease under different population contexts in a precision medicine framework illustrates this point in Chap. 8.

Also in Chap. 8, the analysis of Bateson-Dobzhansky-Müller incompatibilities illustrates the use of the models provided in this book to analyze specific details of historical hypothesis proposed in the context of theoretical evolutionary biology. Since the two aforementioned cases imply two factors—either one gene and one environment or two genes, respectively—the ARNOIA upgrade of NOIA developed in Chap. 6 had to be applied in order to ensure full orthogonality. To end up with, one more example dealt with in Chap. 8—the analysis of the *ACP1* human polymorphism in Europe—illustrates a creative use the models in question enable, particularly in virtue of their accuracy. In particular, they can be used to infer fitness values (i.e., a genotype-to-phenotype map in which the phenotype is fitness itself) from equilibrium frequencies.

In this chapter, we begin by reviewing the use of the top-down and of the bottom-up approaches to quantitative genetics. Then, we look upon NOIA in the light of several lessons learnt from the history of quantitative genetics and evolutionary biology (which were raised in Chap. 1). All of this enables us to deepen into existing inertias related to the black box perspective in quantitative genetics. Finally, the issue of a more fruitful coalition of quantitative genetics and evolutionary biology—already raised in the previous chapter—is here resumed.

9.2 Ups and Downs

Each new scientific discovery is likely to raise overblown expectations. This seems to have been the case of quantitative trait locus (QTL) analysis (as already pointed out by Álvarez-Castro 2020). The basic principle of gene mapping was outlined in the 1960s (Thoday 1961), but it was the work of Lander and Botstein (1989) that opened the door to inspecting genetic architectures underlying quantitative traits of interest in practice. This new perspective was labeled as a bottom-up approach, in contrast to the alternative top-down strategy followed by quantitative genetics hitherto—i.e., the one discussed above, which enabled applications based on variance decompositions whilst remaining parsimonious about the underlying responsible genes (Zuk et al. 2012). It seems understandable that such a paradigm shift aroused scientific excitement.

Although the scope of QTL analysis was initially suitable for model (or at least very easy to manipulate) species, additional developments were proposed to broaden its applicability (e.g., Mott et al. 2000; Slate 2005; Besnier et al. 2010). Moreover, gene mapping went beyond QTL analysis by way of genome-wide association studies (GWAS). The basic principle of GWAS was outlined already in the 1970s in relation to the study of human diseases (Bodmer 1973; McDevitt and Bodmer 1974) but only proved to be feasible in practice in the 2000s (Ozaki et al. 2002). Indeed, it turned particularly consequential in combination with the International HapMap project (International_HapMap_Consortium 2003). In perspective, about two decades of gene mapping can be found to have entrenched the new bottom-up strategy in the corpus of quantitative genetics (see additionaly Rifkin 2012).

9.2.1 Gene Mapping Questioned

However, satisfaction must be measured relative to the expectations previously raised, which may be claimed to have been too demanding at least from some perspectives (see, e.g., Flint and Mott 2001). Indeed, in the context of using QTL analyses for animal and plant breeding, the following excerpt reflects utter disappointment: "The availability of marker panels of thousands of SNPs does [...] appear to be bringing in a real paradigm shift following the pioneering study of Meuwissen et al. (2001), and seems likely to be less of a false dawn than the use of individual QTL" (Hill 2010). More to the point, in the same year, the same highly influential researcher referred to the World Congress on Genetics Applied to Livestock Production (in the speech he gave at the closing ceremony of its 10th edition) as "the world congress of genomic selection applied to livestock production."

The aforementioned criticism on QTL analysis can be considered to have been cast in general upon gene mapping, as more recently also GWAS has been hit by disappointment—being in particular expressed as the (at least relative) "failure of GWAS" (Teperino 2020). In any case, two facts are of particular importance to us at this stage. First, expectations about the time schedule of novel scientific proposals to bring new practical tools of standard use should be made cautiously, at the cost of generating disappointment in the mid or even short term. And second, disappointment on former scientific proposals may be easily reinforced by the excitement of newer ones.

9.2.2 The Time of Genomic Selection

Genomic selection made the focus of quantitative genetics to oscillate back towards the top-down approach—it upgraded the black box perspective of quantitative genetics to the molecular era. Indeed, with genomic selection, individuals are selected based on oversized sets of marker genotypes. These markers include many with very little chances of having any influence on the trait at all, in order to higher the chances of accounting amongst them for most of the ones actually improving selection response. In accordance with the infinitesimal model, only the additive effects of the markers were originally implemented in the analyses, which already implies considerable computational demands.

Nevertheless, implementing interaction effects (both dominance and epistasis) in the analyses was eventually proposed to improve the selection of the genomic panels—particularly, using NOIA (Chevalet et al. 2010; Toro and Varona 2010; Wittenburg et al. 2011). Indeed, the pendulum of quantitative genetics did not last too much in changing course from the original top-down approach of genomic selection towards a more bottom-up focused one. The implementation of interactions with NOIA in genomic selection by Vitezica et al. (2017), particularly enabling to properly account for departures from the Hardy-Weinberg proportions, has been followed by many applications from then on up to the publication of the present book (e.g., Vitezica et al. 2018; Pegot-Espagnet et al. 2019; Joshi et al. 2020; Yadav et al. 2021; Raffo et al. 2022). Just as an example of the achievements of this line of work, accounting for additive-by-additive epistasis with NOIA has been found to improve prediction ability in a 16.5% (Raffo et al. 2022).

NOIA has also been used to show that epistasis may be a major cause of the inefficiency of transferring results of genomic prediction across populations and to estimate genetic correlation between purebred and crossbred performance from information obtained in purebred parental lines only (Duenk et al. 2020, 2021). Implementing epistasis with NOIA has also enabled simulations showing that phenotypic selection may outperform genomic selection in selection response over 50 generations and to maintain a higher genetic variance (Wientjes et al. 2022). In short, it can be said that genomic selection is already called into question in virtue of potential structural limitations.

9.2.3 Finding Our Way Out of the Spiral

In order to escape the trick of being dragged once and again into cycles of excitement and disappointment, it seems sensible to look back and take advantage of the experience gained for planning the next move. On the one hand, NOIA was developed to account for complex genetic architectures and population facts so that the parsimonious assumption of the relatively simple infinitesimal model could be tested. NOIA enables to overcome serious limitations of gene-mapping techniques and novel ways to interpret and analyze the obtained genetic architectures (see, e.g., the two previous chapters of this book).

On the other hand, genomic selection was promoted as an alternative to gene mapping, particularly in the context of animal and plant breeding and in line with the infinitesimal model. Thus, genomic selection revitalized the black box perspective of quantitative genetics with the strength of dense panels of molecular markers. Paradoxically, however, NOIA has made it possible not only to reveal limitations of genomic selection but also to improve it, as mentioned above. Furthermore, this situation may look even more paradoxical as the routine application of NOIA in gene-mapping procedures has not yet been seriously undertaken.

Overall, even the most exciting research tools may take time to be tuned, to be applied properly, and to eventually lead to consequential findings. Dismissing a line of work whose neon lights seem to fade away, in favor of the new flashing neon lights of an alternative proposal, may make us miss or severely delay the achievement of the best each line of work may be capable to proffer. Furthermore, lines of work that initially seem to oppose each other may become less conflicting over time, in one or another way. NOIA has proven to be useful in the context of the top-down approach and it is here argued that its potential to aid its original purpose—in the context of the bottom-up approach—should also be harnessed. In what follows, we delve into this point from a historical perspective.

9.3 Drawing on Mendel's Experience

In Chap. 1, we have analyzed the way serious difficulties faced to start the engine of genetics research were eventually overcome. We have there noted that not only the proper foundation of genetics but also that of quantitative and population genetics after it are thought to have been significantly delayed due to those hurdles. Consequently, we have pointed out some mistakes that we should be able not to repeat, at the risk of continuing to hinder the progress of genetics research and, thus, the return of applications of practical use to the society that sustains it. We shall hereafter try to apply those lessons to our current juncture.

Two of the lessons we have stressed in Chap. 1 were related with the difficulties of the researches of the time to acknowledge and follow up on the contribution of Mendel. Indeed, Mendel's work was initially overlooked and had to be awaken about 30 years later for genetics to be born as a scientific field. The first lesson dealt with the need of fairly assessing breakthrough proposals and the second one was about the danger of dismissing genetic interactions. We shall first resume the first lesson.

9.3.1 Take Innovative Proposals Seriously

It should be thought twice before dismissing proposals that may originally seem to swim upstream, as Mendel's findings themselves did one and a half centuries ago. In this regard, the main objective of this book may be expressed as providing theory enabling new practical applications of advanced genetic modeling (particularly, NOIA) in the fields of quantitative genetics and evolutionary biology. It is understandable that researchers need to be reassured that advanced genetic modeling pays off in relation to just applying the conventional models (e.g., the F_2 and the F_∞ models, presented in Chap. 3), as this may imply a nontrivial increase in effort—training, time, and research resources in general. This is why the Addendum of this book shows NOIA to meet the standards of measurement theory. Under the present heading, we probe along the same lines by illustrating that many key scientific goals cannot be achieved in practice unless a bottom-up approach with at least a certain degree of advanced genetic modeling is applied.

First, let us briefly recall the use of NOIA in genomic selection. Assuming the infinitesimal model, it is possible to hypothesize why the association between alleles favoring selection response and a panel of selected markers vanishes over generations. In fact, recombination appears conceptually as a simple and realistic hypothesis. However, as mentioned above, when analyzing the case in more detail, epistasis has been found as a much more plausible explanation. Thus, more complex genetic modeling than the infinitesimal model was necessary for assessing and actually improving the performance of genomic selection in practice.

In perspective, the infinitesimal model was found to be a potential explanation to many historical observations. One striking phenomenon that was observed in artificial selection experiments is the reaching of a plateau followed by a new burst of selection response, which can be explained by exhaustion of genetic variability followed by the appearance of new mutations (see, e.g., Mackay et al. 1994). However, alternative explanations exist that can now be tested using the bottom-up approach. For instance, a line of artificial selection for growth rate in chicken (Carlborg et al. 2006) displays a plateau followed by additional selection response that does not match the only mutation that has been found to occur during the selection process (Pettersson et al. 2013; see also Le Rouzic and Álvarez-Castro 2016). Analogous to what has been discussed about genomic selection above, strong support has been found for that plateau to have been caused by epistasis (Le Rouzic and Álvarez-Castro 2016). Naturally, it was only by using more advanced genetic modeling than the infinitesimal model itself that attaining this result was possible. In this case, it was enough to use Zeng et al. (2005) G2A model (Expressions 3.7 and 3.8) as modified by Álvarez-Castro and Carlborg (2007; see Expressions 5.11 and 5.12 and related text in Chap. 5).

Some other analyses have been made applying advanced genetic modeling—particularly, the original setting of NOIA (Álvarez-Castro and Carlborg 2007; see again Expressions 5.11 and 5.12)—to the aforementioned chicken dataset, of a cross between wild and domesticated chicken. This way, for instance, epistasis has been found to underlie transgressive segregation (Álvarez-Castro et al. 2012), and the way growth rate is temporized (Le Rouzic et al. 2008). These two works were carried out in virtue of the genetic filtering procedure, which could be applied using NOIA, as described in Chap. 7.

In all these analyses of the bottom-up approach with advanced genetic modeling, the models of genetic effects used approach orthogonality in what regards several particularities of the data analyzed (e.g. different allele frequencies), but not all possible ones (e.g. departures from linkage equilibrium). As noted by Vitezica et al. (2017), approaching orthogonality already results in a higher reliability of the results of the analyses, even when full orthogonality is not achieved. Thus, analyses in which full orthogonality is reached make an even clearer point in what regards the convenience of advanced genetic models of genetic effects. Three such cases are dealt with in Chap. 8 and summarized in the introduction above (and thus will not be repeated again here).

All in all, we have gone through a number of cases illustrating that applying advanced genetic modeling is necessary for unraveling evolutionary phenomena of practical interest. Thus, even when implying certain demands in terms both theoretically and experimentally, advanced genetic modeling in general and NOIA in particular should not be dismissed, to the cost of imposing severe limitations to quantitative genetics research. As discussed in Chap. 1, analogous such drawbacks are regrettable to have already occurred in the past, having led to significant delays in scientific progress. Furthermore, as advanced genetic modeling is necessary to properly analyze interactions in the study of biological inheritance, with the above, we have also partly covered the second lesson from Mendel we derived in Chap. 1—which we cover in more detail hereafter.

9.3.2 Considering Interactions May Really Aid

The concept of dominance enabled Mendel to interpret the proportions of phenotypes resulted from his crosses as the interaction between "factors" inherited from the parents. The principle of segregation could then be sustained by this finding. Thus, the central role of interactions in the discovery of the most basic principles of genetics is just out of question. Despite that, a long-lasting debate soon emerged about how much attention should be paid to interactions, particularly to epistasis (see Chap. 6). In this context, very prestigious researchers have aligned with Fisher's dismissal of the effects of epistasis on the response to selection in recent times (e.g., Hill et al. 2008; Crow 2010). However, what we have discussed under the previous heading just confirms that interactions, whether elusive to unravel, repeatedly prove not worth shunning (see also Zuk et al. 2012; Hansen 2013).

It is known since long that looking for interactions may disclose loci that could not stick out for their additive effects. For instance, two-dimensional QTL scans have disclosed pairs of epistatic loci that could not be found for their marginal effects through one-dimensional scans (e.g., Carlborg and Andersson 2002; Carlborg and Haley 2004). In spite of that, the paradigm that interactions should be inspected only for loci that have already been found to have significant additive effects is not straightforward to overcome (see, e.g., Malosetti et al. 2013). Reassuringly, interactions may reveal loci that do not stick out for their additive contributions also in the context of GWAS (for a recent example see Reverter et al. 2020).

Nevertheless, it remains true that making practical use of the results of mapping experiments may sometimes turn elusive, particularly due to the difficulty of efficiently disentangling complex genetic architectures in which interactions are present. As explained above, some researchers seem to have gotten frustrated about having to deal with this—particularly in what regards animal and plant breeding. But then, it is remarkable that when genomic selection emerges, precisely in the context of animal and plant breeding, as an alternative to once again go around the putative complexity of the real underlying genetics of a trait (i.e., the bottom-up approach), implementing interactions to the analyses eventually (indeed, fairly soon) turns out to be convenient, to say the least.

9.4 Two Remaining Lessons Not to Miss

We have so far taken advantage of historical knowledge on Mendel's work, particularly about the importance of interactions and about the risk of not properly evaluating the potential impact of a scientific proposal. In accordance with the latter issue, the theoretical developments provided in this book under the label of NOIA are challenged from the perspective of measurement theory in the Addendum. As pointed out in Chap. 1, we would still like to view NOIA in light of two additional lessons the history of genetics teaches us. We shall first consider the problem of personality clashes and then, from a more positive perspective, analyze the current juncture of advanced models of genetic effects in search of the most solid ground from which to increase scientific return to society.

9.4.1 Bypass Personality Clashes

The dispute between Mendelians and biometricians after the so-called rediscovery of Mendel's work is assumed to be the major cause of a delay of 15 years in scientific progress in genetics (Provine 1971; see Chap. 1 of this book). Mendelians honored their label by advocating Mendelian inheritance of meristic characters as a major force in evolution. Biometricians interpreted evolution as a continuous process, i.e., in the sense that Darwin understood gradualism. Both sides made the same mistakes—to presuppose that Mendel's work would not typically lead to Darwin's gradual evolution and that Mendelian inheritance would be incompatible with selection bringing a trait beyond its observed range. Then, personality clashes precluded a reconciliation between the two parties.

Although Fisher's (1918) genetic variance decomposition is acknowledged to have completely gotten around the original dispute between Mendelians and biometricians, it can equally be said to have perpetuated personality clashes and its consequences within the field of genetics. Indeed, Fisher's (1918, 1930) intuitive dismissal of the implications of epistasis in relation with the action of selection crashed against the relevance the Mendelians gave to it. The term epistasis was coined by Bateson (1907), the leading Mendelian who had 50 years earlier also coined not less than the term "genetics" (Bateson and Saunders 1902). Fisher opted for the infinitesimal model both as a tribute to Charles Darwin and a condemnation to Bateson. Fisher embraced the infinitesimal model as a form of "neo-Darwinian gradualism" in line with advice from his mentor, even knowing that he (Charles Darwin's son Leonard Darwin) eventually stepped back from that view (Stoltzfus and Cable 2014).

Later on, also Wright (1932; see also Provine 1986) stood out in the defense of epistasis, particularly as a crucial conditioning factor of the evolutionary process from a population genetics perspective, as already pointed out in Chap. 3. In Chap. 6, we have likewise mentioned that that such controversy resurfaced with the advent of QTL analysis. Indeed, while it seems sensible to realize that genetic mapping has serious limitations if interactions—particularly, epistasis—are not considered, some researchers have construed that considering interactions appropriately will not either allow us to overcome such limitations, as mentioned above.

It is possible to interpret the above as two threads that have been followed in parallel for a long time. On the one hand, there are conceptual connections between the focus on Mendel's results by the Mendelians, identifying epistasis as a powerful explanatory population genetics mechanism by Wright and assuming gene-mapping as a procedure to unravel realistically complex genetic architectures, for all of them fit to a bottom-up perspective. On the other hand, we can in general terms identify a top-down perspective in focusing on gradual variation and evolution by the biometricians. Indeed, they eventually used the decomposition of the genetic variance and the infinitesimal model as practical quantitative genetics tools to perform empirical research based on the work of Fisher. That way, they remained unaware of the genetic basis of a trait when harnessing molecular genetics in animal and plant breeding by means of genomic selection.

Einstein's metaphoric claim that God does not play dice reflects the difficulties he had in valuing the working perspective of quantum mechanics. Singh (2015) finds parallelisms between Darwin and Einstein and between Mendel and quantum mechanics and advocates that—analogous to what occurred to Einstein—a fictitious encounter between Darwin and Mendel would not raise interesting scientific synergies. However, it remains true not only that Mendel foresaw that his findings would be compatible with gradual variation, as already pointed out in Chap. 1, but also that Mendel thought that something was still lacking in Darwin's theory of evolution and that his work could potentially fill that gap as Singh (2015) himself mentions.

Then, in what regards the two threads mentioned above, we may position Darwin's concern on the evolution of phenotypes in the top-down thread. Indeed, he can be understood as having worked under a black box perspective in what regards a theory of biological inheritance, as his pangenesis theory was refuted—incidentally, by his own cousin, Galton (see, e.g., Liu 2008). On the other hand, Mendel found the key to the black box of biological inheritance. Whether he sufficiently valued Darwin's works or not, his efforts in making his scientific work understood and accepted—and his professional duties—did not leave him much time to develop further his intuitions about gradual variation and the theory of evolution by natural selection.

Overall, many different circumstances may slow down the pace of science. The case of Galton, a major inspiration for the biometricians who nevertheless adhered to the Mendelians (as pointed out in Chap. 1; see Provine 1971), further illustrates both that affinities (like the ones discussed right above) must be done with caution and how difficult it can get to be to properly navigate the crossroads of biological inheritance. Ultimately, we may remain doubtful about the possibilities that Darwin and Mendel could have made significant scientific leaps forward through a fictitious encounter, but we would hope or at least wish that personal issues would not have interfered with their capability to go beyond the inertia of their initial research paradigms.

In what regards the top-down and the bottom-up approaches here discussed, it seems just as reasonable to wish our current leading scientists not to take research results personality, which may open the way for enriched evolutionary quantitative genetics. And yet, revising once again the historical records, it may look more sensible to place hope for a consensus on those considered to be "middle-level" researchers. Indeed, Kim (1994) provides evidence that it is among these that are the ones who most of the times actually tackle the tasks needed for a new idea to become accepted into the core knowledge of a research field. By acting as critics, experimenters, reviewers, and validators, they are able to bypass personality clashes and thus rise above biases that the scientific elite gets stuck with.

9.4.2 Harness a Favorable Scenario for a Consensus

We hereafter propose NOIA as a valuable tool in seeking a consensus on the controversy between the top-down and the bottom-up approaches. Indeed, it can aid in neutralizing biases inherited from the past. In Chap. 1, we have shown that various empirical results and theoretical developments had eroded the prejudice that Mendelian inheritance would not typically underlie continuous variation—and, thus, gradual evolution—before Fisher's (1918) proposal on this subject was published and generally accepted. However, the viewpoints of the parties remained controversial, particularly in what regards the role of epistasis in the action of selection, as explained above.

The top-down approach circumvented epistasis through the use of the infinitesimal model, as discussed in Chaps. 2, 3, and 6 and above. As a simplistic assumption in the lack of tools to inspect more complex, real genetic architectures, the use of the infinitesimal model fits to a black box perspective that, as such, would not be supposed to necessarily explain all possible empirical observations. Indeed, although the infinitesimal model is not less than "commonly considered to be the founding principle of quantitative genetics" (Crouch and Bodmer 2020), it is known to lead to predictions that can be used to call itself into question. One example of this is an expected increase of the selection limit through selection within families, as compared to mass selection (Dempfle 1974), which is in contradiction with empirical observations (Gallego and López-Fanjul 1983; see also Caballero and Hill 1992).

In other words, using the infinitesimal model in the context of a black box strategy may be useful in many cases, which does not make it realistic. Or, as expressed by Hill (2014), the model "does not have to be true to be useful." For a long time (in the absence of methods to inspect real genetic architectures), using a black box that sacrifices realism simply was not a choice. But, as that of metaphor (see, e.g., Smith 2009), the price of the use of assumedly untrue models is eternal vigilance. Keeping in mind that the infinitesimal model is not true implies eternal vigilance over generality and precision in practice. Indeed, it is one thing that generality and precision are the two upsides of these type of models themselves (meaning type II models; see Addendum) and quite another not to pay sufficient attention to whether and how these models may enable practical applications with real data. Some of the above concerns about the use of heritability illustrate that not all applications of the top-down approach have been vigilant enough about this.

When the bottom-up approach of gene-mapping came into play, the possibility of filling gaps left by the top-down approaches emerged. Deciphering genetic architectures (i.e., looking inside the black box at last) could finally enable researchers to evaluate empirically many theoretical and/or conceptual proposals. Among these proposals, the analysis of the role of epistasis in evolution was one with particular historical relevance, as mentioned above. Wright (1932) envisaged epistasis as a major mechanism shaping adaptation in natural populations, as already pointed out in Chaps. 3 and 6.

If epistasis has a high explanatory potential for the evolution of natural populations, there is no reason to believe that it is not conditioning the outcome of artificial selection programs just as well. Besides, the explanatory power of epistasis goes beyond Wright's shifting-balance theory. The Bateson-Dobzhansky-Müler model of speciation dealt with in Chap. 8 also helps us understand why analyzing epistasis has long been deemed crucial. Epistasis has also been found as a mechanism ultimately explaining the maintenance of the Drosophila inversion polymorphisms. In the context of this line of work undertaken by population geneticists since the 1940s, it is worth noting that Hedrick and Murray (1983) did "not want to be right for the wrong reason," which is opposed to looking for useful untrue models—and in fact matching conceptually the bottom-up approach.

But soundly unveiling right reasons must also have a high price, which surely includes a great deal of patience. This cost seems not to have been sufficiently taken into account, as the above interpretations of a "false dawn" of QTL analysis and of a "failure of GWAS" suggest. In particular, before throwing in the towel for not achieving convincing enough results (despite employing expensive technologies in the fields of molecular genetics, phenotyping, biostatistics, computational resources, and bioinformatics), it may be sensible to give it another try with the greased machinery of advanced genetic modeling.

Overall, there is currently a scenario where NOIA can help reach consensus by assisting both bottom-up and top-down approaches, thus generating a synergy that contributes to the progress of evolutionary quantitative genetics. Mendel and Fisher's works provided the key to unlocking the basis of biological inheritance and for applying it to the study of quantitative traits, respectively. NOIA generalizes and properly unifies the study of biological inheritance both at the individual and at the population levels. With this, NOIA provides the key to reliably opening the black box of genetic architectures in empirical studies. Incidentally, restricting ourselves to a purely black box perspective, while we have a key that could open it, evokes the Stockholm syndrome. This fact is all the more curious since the key we can use to overcome this syndrome was forged—i.e., the original version of NOIA was submitted—barely 60 kilometers from that city (Álvarez-Castro and Carlborg 2007).

9.5 Closing Perspective

A general theory of genetic effects (a broad GP map) is provided in this book under the name of NOIA. Only if geneticists tackle the task of carrying on the use of advanced genetic modeling for deciphering and analyzing genetic architectures, time will reveal the real extent to which gene mapping experiments may overcome its current limitations. Results achieved with different installments of NOIA in bottom-up approaches (as the ones illustrated in the previous chapter) reassure us that devoting further efforts in this direction will be rewarding. Moreover, we have also shown above that implementing NOIA to improve and assess an originally intended top-down approach (genomic selection) adds to this view. We have thus shown that NOIA contributes to merge top-down conventional quantitative genetics procedures with bottom-up oriented population genetics ones into an inclusive body of evolutionary quantitative genetics (see also Álvarez-Castro 2016).

Ultimately, the progress of evolutionary quantitative genetics is particularly dependent on a clear-cut and applicable distinction between the "biological and the statistical effects" (Hansen 2023), i.e., between the genetic (and environmental) effects at the individual and at the population levels. The current state of development of the NOIA model, as provided in this book, enables such distinction in the general scope—both in what regards the complex genetic architectures and the broad population conditions that can now be considered in the analyses. NOIA can thus be harnessed to overcome existing critical limitations of evolutionary quantitative genetics, as has recently been made explicit, for instance, in relation to the study of evolvability (Houle et al. 2023).

References

Álvarez-Castro JM (2016) Genetic architecture. In: Wolf JB (ed) Encyclopedia of evolutionary biology. Oxford Academic Press, Oxford, pp 127–135

Álvarez-Castro JM (2020) Gene-environment interaction in the era of precision medicine - filling the potholes rather than starting to build a new road. Front Genet 11:921

Álvarez-Castro JM, Carlborg Ö (2007) A unified model for functional and statistical epistasis and its application in quantitative trait loci analysis. Genetics 176:1151–1167

Álvarez-Castro JM, Yang RC (eds) (2015) Models and estimation of genetic effects. Frontiers Media, Lausanne

Álvarez-Castro JM, Le Rouzic A, Andersson L, Siegel PB, Carlborg O (2012) Modelling of genetic interactions improves prediction of hybrid patterns—a case study in domestic fowl. Genet Res (Camb) 94:255–266

Avery OT, Macleod CM, Mccarty M (1944) Studies on the chemical nature of the substance inducing transformation of pneumococcal types nducton of transformation by a desoxyribonucleic acid fraction isolated from pneumococcus type ii. J Exp Med 79:137–158

Bateson W (1907) Facts limiting the theory of heredity. Science 26:649–660

Bateson W, Saunders ER (1902) The facts of heredity in the light of Mendel's rediscovery. In: Bateson W (ed) Report of the evolution Committee of the Royal Society of London, pp 87–160

Besnier F, Le Rouzic A, Álvarez-Castro JM (2010) Applying QTL analysis to conservation genetics. Conserv Genet 11:399–408

Bodmer WF (1973) Genetic factors in Hodgkin's disease: association with a disease-susceptibility locus (DSA) in the HL-A region. Natl Cancer Inst Monogr 36:127–134

Bulmer MG (1980) The mathematical theory of quantitative genetics. Oxford University Press, Oxford

Caballero A, Hill W (1992) Artificial selection responses. Annu Rev Ecol Syst 23

Carlborg Ö, Andersson L (2002) Use of randomization testing to detect multiple epistatic QTLs. Genet Res 79:175–184

Carlborg Ö, Haley CS (2004) Epistasis: too often neglected in complex trait studies? Nat Rev Genet 5:618–625

Carlborg Ö, Jacobsson L, Ahgren P, Siegel P, Andersson L (2006) Epistasis and the release of genetic variation during long-term selection. Nat Genet 38:418–420

Chevalet C, Servin B, Sancristobal M (2010) Including non-additive effects in Bayesian methods for the prediction of genetic values from genome-wide SNP data. In: 10th WCGALP. German Society of Animal Science

Cheverud JM, Routman EJ (1995) Epistasis and its contribution to genetic variance components. Genetics 139:1455–1461

Crouch DJM, Bodmer WF (2020) Polygenic inheritance, GWAS, polygenic risk scores, and the search for functional variants. Proc Natl Acad Sci U S A 117:18924–18933

Crow JF (2010) On epistasis: why it is unimportant in polygenic directional selection. Philos Trans R Soc Lond B Biol Sci 365:1241–1244

Dempfle L (1974) A note on increasing the limit of selection through selection within families. Genet Res 24:127–135

Duenk P, Bijma P, Calus MPL, Wientjes YCJ, Van Der Werf JHJ (2020) The impact of non-additive effects on the genetic correlation between populations. G3 10:783–795

Duenk P, Bijma P, Wientjes YCJ, Calus MPL (2021) Predicting the purebred-crossbred genetic correlation from the genetic variance components in the parental lines. Genet Sel Evol 53:10

Fisher RA (1918) The correlation between relatives on the supposition of Mendelian inheritance. Trans Roy Soc Edinburgh 52:339–433

Fisher RA (1930) The genetical theory of natural selection. Clarendon, Oxford

Flint J, Mott R (2001) Finding the molecular basis of quantitative traits: successes and pitfalls. Nat Rev Genet 2:437–445

Franklin R, Gosling RG (1953) Molecular configuration in sodium thymonucleate. Nature 171: 740–741

Gallego A, López-Fanjul C (1983) The number of loci affecting a quantitative trait in Drosophila melanogaster revealed by artificial selection. Genet Res 42:137–149

Hansen TF (2013) Why epistasis is important for selection and adaptation. Evolution 67:3501–3511

Hansen TF (2023) Variation, inheritance, and evolution: a primer on evolutionary quantitative genetics. In: Hansen TF, Houle D, Pavlicev M, Pélabon C (eds) Evolvability: a unifying concept in evolutionary biology? The MIT Press, Cambridge, pp 73–100

Hansen TF, Wagner GP (2001) Modeling genetic architecture: a multilinear theory of gene interaction. Theor Popul Biol 59:61–86

Hedrick PW, Murray E (1983) Selection and measures of fitness. In: Ashburmer M, Carson H, Thompson JN Jr (eds) The genetics and biology of drosophila. Academic Press, London

Hershey AD, Chase M (1952) Independent functions of viral protein and nucleic acid in growth of bacteriophage. J Gen Physiol 36:39–56

Hill WG (2010) Understanding and using quantitative genetic variation. Philos Trans R Soc Lond B Biol Sci 365:73–85

Hill WG (2014) Applications of population genetics to animal breeding, from Wright, fisher and lush to genomic prediction. Genetics 196:1–16

Hill WG, Goddard ME, Visscher PM (2008) Data and theory point to mainly additive genetic variance for complex traits. PLoS Genet 4:e1000008

Houle D, Pélabon C, Pavlicev M, Hansen TF (2023) Conclusion: is evolvability a new and unifying concept? In: Hansen TF, Houle D, Pavlicev M, Pélabon C (eds) Evolvability: a unifying concept in evolutionary biology? The MIT Press, Cambridge, pp 373–388

International_Hapmap_Consortium (2003) The international HapMap project. Nature 426:789–796

Joshi R, Meuwissen THE, Woolliams JA, Gjoen HM (2020) Genomic dissection of maternal, additive and non-additive genetic effects for growth and carcass traits in Nile tilapia. Genet Sel Evol 52:1

Kim K-M (1994) Explaining scientific consensus: the case of Mendelian genetics. Guilford Press, New York

Lander ES, Botstein D (1989) Mapping mendelian factors underlying quantitative traits using RFLP linkage maps. Genetics 121:185–199

Le Rouzic A, Álvarez-Castro JM (2016) Epistasis-induced evolutionary plateaus in selection responses. Am Nat 188:E134–E150

Le Rouzic A, Álvarez-Castro JM, Carlborg Ö (2008) Dissection of the genetic architecture of body weight in chicken reveals the impact of epistasis on domestication traits. Genetics 179:1591–1599

Liu Y (2008) A new perspective on Darwin's pangenesis. Biol Rev Camb Philos Soc 83:141–149

Mackay TF, Fry JD, Lyman RF, Nuzhdin SV (1994) Polygenic mutation in Drosophila melanogaster: estimates from response to selection of inbred strains. Genetics 136:937–951

Malosetti M, Ribaut JM, Van Eeuwijk FA (2013) The statistical analysis of multi-environment data: modeling genotype-by-environment interaction and its genetic basis. Front Physiol 4:44

Mcdevitt HO, Bodmer WF (1974) HL-A, immune-response genes, and disease. Lancet 1:1269–1275

Meuwissen TH, Hayes BJ, Goddard ME (2001) Prediction of total genetic value using genome-wide dense marker maps. Genetics 157:1819–1829

Mott R, Talbot CJ, Turri MG, Collins AC, Flint J (2000) A method for fine mapping quantitative trait loci in outbred animal stocks. Proc Natl Acad Sci U S A 97:12649–12654

Ozaki K, Ohnishi Y, Iida A, Sekine A, Yamada R, Tsunoda T, Sato H, Sato H, Hori M, Nakamura Y, Tanaka T (2002) Functional SNPs in the lymphotoxin-alpha gene that are associated with susceptibility to myocardial infarction. Nat Genet 32:650–654

Pegot-Espagnet P, Guillaume O, Desprez B, Devaux B, Devaux P, Henry K, Henry N, Willems G, Goudemand E, Mangin B (2019) Discovery of interesting new polymorphisms in a sugar beet (elite × exotic) progeny by comparison with an elite panel. Theor Appl Genet 132:3063–3078

Pettersson ME, Johansson AM, Siegel PB, Carlborg O (2013) Dynamics of adaptive alleles in divergently selected body weight lines of chickens. G3 3:2305–2312

Provine WB (1971) The origins of theoretical population genetics. University of Chicago Press, Chicago

Provine WB (1986) Sewall Wright and Evolutionary biology. University of Chicago Press, Chicago

Raffo MA, Sarup P, Guo X, Liu H, Andersen JR, Orabi J, Jahoor A, Jensen J (2022) Improvement of genomic prediction in advanced wheat breeding lines by including additive-by-additive epistasis. Theor Appl Genet 135:965–978

Reverter A, Vitezica ZG, Naval-Sanchez M, Henshall J, Raidan FSS, Li Y, Meyer K, Hudson NJ, Porto-Neto LR, Legarra A (2020) Association analysis of loci implied in "buffering" epistasis. J Anim Sci 98

Rifkin S (ed) (2012) Quantitative trait loci (QTL). Springer, New York

Singh RS (2015) Limits of imagination: the 150th anniversary of Mendel's Laws, and why Mendel failed to see the importance of his discovery for Darwin's theory of evolution. Genome 58:415–421

Slate J (2005) Quantitative trait locus mapping in natural populations: progress, caveats and future directions. Mol Ecol 14:363–379

Smith B (2009) The price of metaphor is eternal vigilance: language metaphors in popular genetics. Int J Humanit 6:79–86

Stoltzfus A, Cable K (2014) Mendelian-Mutationism: the forgotten evolutionary synthesis. J Hist Biol 47:501–546
Teperino R (2020) Preface. In: Teperino R (ed) Beyond our genes. Patophysiology of gene and environment ineraction and epigenetic inheritance. Springer, Cham
Thoday JM (1961) Location of polygenes. Nature 191:368–370
Toro MA, Varona L (2010) A note on mate allocation for dominance handling in genomic selection. Genet Sel Evol 42:33
Vitezica ZG, Legarra A, Toro MA, Varona L (2017) Orthogonal estimates of variances for additive, dominance, and epistatic effects in populations. Genetics 206:1297–1307
Vitezica ZG, Reverter A, Herring W, Legarra A (2018) Dominance and epistatic genetic variances for litter size in pigs using genomic models. Genet Sel Evol 50:71
Watson JD, Crick FHC (1953) A structure for deoxyribose nucleic acid. Nature 171:737–738
Wientjes YCJ, Bijma P, Calus MPL, Zwaan BJ, Vitezica ZG, Van Den Heuvel J (2022) The long-term effects of genomic selection: 1. Response to selection, additive genetic variance, and genetic architecture. Genet Sel Evol 54:19
Wittenburg D, Melzer N, Reinsch N (2011) Including non-additive genetic effects in Bayesian methods for the prediction of genetic values based on genome-wide markers. BMC Genet 12:74
Wright S (1932) The roles of mutation, inbreeding, crossbreeding and selection in evolution. In: 6th international congress of genetics, pp 356–366
Wu R (1972) Nucleotide sequence analysis of DNA. Nat New Biol 236:198–200
Yadav S, Wei X, Joyce P, Atkin F, Deomano E, Sun Y, Nguyen LT, Ross EM, Cavallaro T, Aitken KS, Hayes BJ, Voss-Fels KP (2021) Improved genomic prediction of clonal performance in sugarcane by exploiting non-additive genetic effects. Theor Appl Genet 134:2235–2252
Yang R-C (2004) Epistasis of quantitative trait loci under different gene action models. Genetics 167:1493–1505
Zeng ZB, Wang T, Zou W (2005) Modeling quantitative trait loci and interpretation of models. Genetics 169:1711–1725
Zuk O, Hechter E, Sunyaev SR, Lander ES (2012) The mystery of missing heritability: genetic interactions create phantom heritability. Proc Natl Acad Sci U S A 109:1193–1198

Check for updates

Addendum: An Acid Test for NOIA 10

Abstract

Measurement theory is nuclear to mathematical modeling. Generally speaking, modeling can be seen as developing mathematical functions (or maps) whose variables and images (the result of applying the functions to particular values of the variables) reflect properties of the subject under study. Those functions enable predictions that can then be compared against empirical observations. Modeling is often undertaken intuitively. Measurement theory provides rules that prevent researchers from falling into modeling errors leading to wrong predictions. In this section, we review common errors made when developing mathematical models within the field of biology and we verify that NOIA is not affected by any of them. In brief, we show that NOIA conforms to the dictates of measurement theory. At the same time, we here provide a practical review of the most important concepts worked out in the previous sections.

10.1 Introduction

In Chap. 1, we have discussed the problems derived from Mendel's work not having been properly pondered by his peers in the nineteenth century. Based on that knowledge, we have in Chap. 9 warned about the risk of not properly pondering advanced genetic modeling (particularly, NOIA) in the twenty-first century. In order to make it easier for researchers to value the mathematical theory of genetic effects presented in this book (advanced genetic modeling in Chaps. 3–7 and NOIA in particular in Chaps. 4–7), we have expounded several applications of it in Chap. 8 and discussed it further in Chap. 9. In this addendum, we add to this evaluation by double-checking the reliability of NOIA from a theoretical perspective. In particular, we here put NOIA to the test in light of the principles of measurement theory.

With this, we show NOIA to be an adequate methodology to analyze genetic effects as they were originally conceived by Fisher (1918; see Chap. 2). In that

J. M. Álvarez-Castro, *Genes, Environments and Interactions*,
https://doi.org/10.1007/978-3-031-41159-5_10

article, Fisher established not less than the merge of Darwin's theory of evolution and Mendel and Galton's studies on biological inheritance of, respectively, discrete and continuous variation. Keeping that in mind, it is not surprising that NOIA is currently key to establish the confluence that underpins the development of evolutionary quantitative genetics (see Chap. 9).

The process of measuring is easily regarded as empirical. Physical measurements of extraordinary historical relevance have been recorded. A remarkable instance of this comes from Eratosthenes of Cyrene, who is acknowledged to find the way to attain a significantly accurate measure of the Earth's circumference around 240 BC (see e.g. Russo 2004 and references therein). Nevertheless, such an empirical realization was only possible on the condition of having previously developed a meticulous theoretical framework—revealing what has to be measured and how to use the measures performed for obtaining the desired information. Making explicit and formalizing the necessary concepts and the rules measurements must fulfill in order to be useful constitutes a field called measurement theory.

10.2 Measurement Theory

In brief, measuring can be defined as adequately extracting meaning from observations, experiments, and models (e.g., Krantz et al. 1971). Measures apply to attributes of objects, organisms, or populations and enable us to assign a quality—often, a numerical one—to a manifestation of such an attribute. Measurement theory elaborates on how the objects of study are assigned to qualities within a particular scale (i.e., on so-called relational structures), in order to guarantee that the former make sense by means of the latter. Measures may be nominal (e.g., species or other taxa), ordinal (integers), or magnitudes. A magnitude is any entity that may be compared (through the relational structure) against a unit. Units may be chosen arbitrarily to some extent but should fulfill invariability, verifiability, and global reach. Any study must be coherent with the type of magnitude in question, which may for instance admit concatenation or not (i.e., they may be extensive or intensive magnitudes, respectively). Coherence is also required with regard of the type of scale the measurement uses to assign a quality—which particularly conditions the transformations of the measurements that are authorized—to the risk of compromising the meaningfulness of the results obtained.

Getting from the abstract foundations of measurement theory (as the ones drafted above) to guidelines for properly measuring genetic effects in practice is not trivial. Fortunately, we can rely here on the intermediate steps that have already been worked out by Houle et al. (2011), who reviewed measurement theory in biology. They discuss general aspects of the application of measurement in biology and conclude that violations of measurement theory are widespread in this research field, thus hindering appropriate scientific progress. As examples, they denounce eight specific practices that are in opposition to the dictates of measurement theory and provide warnings to avoid the undesirable consequences they lead to. Specifically, they emphasize remembering the context, not to use a rubber ruler, interpreting the numbers, respecting scale, not to let statistics overrule meaning, treating measurements as measurements, knowing what the parameters mean, and making meaningful measurements.

Under the next heading, we contrast NOIA with those warnings. To this end, we take advantage of useful insights about theoretical models recalled by Houle et al. (2011). On the one hand, following Levins (1966), they distinguish between three different model strategies. As generality, precision, and realism are three general purposes of theoretical models that can hardly be attained in depth jointly, strategies can be classified in attention to the most sacrificed of those three purposes. Thus, type I models scarify generality in favor of realism and precision, type II models scarify realism in favor of generality and precision, and type III models scarify precision in favor of generality and realism. On the other hand, Houle et al. (2011) point out that Lewontin (1974) conceptualized as dynamically sufficient models those that can lead to predictions whenever their parameters are known, and they would also be empirically sufficient models when their parameters can be estimated with sufficient accuracy in practice as to enable reliable predictions. Theoretical models must be dynamically sufficient in order to meet the standards of measurement theory, although it is not mandatory that they are also empirically sufficient.

10.3 Measurement of Genetic Effects

Houle et al. (2011) regard Fisher's (1918, 1930, 1941) measurements through models of genetic effects as approximately dynamically sufficient, although only in what regards short-term evolution—strictly speaking, under one only episode of selection. Indeed, we have already discussed that measuring variance components (particularly, the additive variance) and thus heritability using a top-down approach involves some delicate assumptions, as discussed in Chaps. 2 and 7 (see also Zuk et al. 2012), which justifies the "approximately" evaluation above. In any case, since measuring phenotypes and relatedness enables predictions of short-term selection in practice, the models are also empirically sufficient.

As explained in Chap. 9, the bottom-up approach was made possible by means of gene-mapping, which makes models of genetic effects to become empirically sufficient also in what regards the genetic effects themselves. With this, models of genetic effects become dynamically sufficient also in what regards long-term selection, since the outcome of selection with specific genetic effects can be predicted. The scope of this approach relies on implementing those models appropriately and, particularly, on achieving generality—the main goal of this book. In any case, it can be considered controversial whether at this juncture gene mapping is also empirically sufficient, as discussed above—although progress in the use of advanced genetic modeling applied to real data cannot be denied, disappointment permeated the field significantly, particularly when it came to use the outcome of mapping experiments for animal and plant breeding.

In what regards that this book provides very general models of genetic effects, thus being able to capture the complexity of real cases, it can now be added that these models may be considered as type III models. It is worth noting, however, that under this conception the lack of precision may reach the level of inoperativeness. In plain words, trying to account for too many realistic parameters in a mapping experiment may lead to not being able to obtain estimates at all, rather than to obtain low-precision estimates. As Houle et al. (2011) stress, measurement theory also dictates to know the limits of the models. Nevertheless, a great advantage of the models in this book is that they can be cut down to any desired particular cases (which would be a mess to develop individually), thus leading to type I models sacrificing generality in favor of realism and precision. In addition, the aforementioned use of the models in this book for predicting the outcome of selection from (putatively non-realistic) specific genetic effects would fall within the context of type II models.

10.3.1 Remember Context

The first of Houle et al. (2011) warning matches several specific features of the measurement of genetic effects in general and of NOIA in particular. To start with, NOIA upgrades genetic models to match the bottom-up context, as mentioned right above. The models used in the first gene-mapping applications were those originally developed under the top-down approach, as soon pointed out by Cheverud and Routman (1995). Currently, NOIA provides a much more explicit modeling of genetic effects with a potential to improve gene mapping and in general interpretation of the meaning and the properties of genetic effects.

More in general, adhering to the top-down approach for several generations of quantitative geneticists directed the research focus towards the additive variance. This might have blurred the crucial fact that not being able to tell apart the interaction variances from the environmental variance does not mean that the former do not exist by themselves or that they have no evolutionary consequences, which is just a small step from assuming the context that epistasis is a rare and/or uninteresting phenomenon (see Chap. 7). As discussed in Chaps. 2 and 9, the top-down context fits a black box strategy under which it is not possible to unveil either epistasis or other features of a complex genetic architecture. But that does not mean that those features do not occur. NOIA (and more in particular, its upgraded variant ARNOIA) enables and has actually been used to account for complex genetic architectures, including epistasis under non-equilibrium population facts.

Another context that is crucial to always remember when dealing with models of genetic effects is that of the selection regime. As most of quantitative genetics was developed with animal and plant breeding in mind, the phenotype itself was understood as an approximation of fitness, as the action of selection is understood to be directional. However, that does not have to always be the case. The study of the *ACP1* human enzyme in a European population in Chap. 8 illustrates that sometimes transcending apparently obvious contexts, such as the one in question, may be key to get out of the maze. One other way of remembering this lesson is to never get distracted from honoring the family of hypotheses under which a work is being carried out, so that an emergency light always flashes when one of them is broken (as also stressed by Houle et al. 2011).

Also through the study of the *ACP1* system in Chap. 8, we have warned about another situation that sometimes gets researches confused about the context at which the analyses performed make sense. In particular, it is crucial to always have the alarm for reductionism connected. Indeed, we have shown that analyzing the three-allele *ACP1* system as a reduced biallelic system (by merging two of the three alleles into a single category) provided misleading results—that the polymorphism was necessarily transient. It was by extending NOIA to multiple alleles that the case could be analyzed in all its complexity—and even estimates of fitness values explaining the maintenance of the polymorphism were obtained.

10.3.2 Do Not Use a Rubber Ruler

We shall here further elaborate on the use of heritability. Evolutionary potential is a central concept in evolutionary biology. Measuring evolvability and being able to compare measures constitutes a typical example of how a relational structure may provide meaning (comparing the evolvability of different traits) from observations (phenotypes and relatedness). Houle et al. (2011)—and more in depth Hansen et al. (2011)—review the use of heritability as a measure of evolvability (see also Hansen 2016). Narrow sense heritability is the ratio of additive and phenotypic variance. It is to be noted, however, that the latter is correlated with the former (the additive variance is a component of the phenotypic variance). Thus, the scale used for comparison across traits and populations is a rubber ruler that in particular depends on the measured attribute (by stretching or shrinking). Technically speaking, the relational structure does not provide the desired meaning.

This is a good example that the violation of the dictates of measurement theory do not only imply that it just would be more appropriate theoretically to do things in other (more difficult) ways but that non-desired practical problems will surely arise as well. Indeed, the above can be expressed in a more practical way as follows. It can be said that the heritability, h^2, seems to be a good index of (at least short term) evolvability because as $R = h^2S$ (Expression 2.28), the higher the heritability the larger the response to selection, R. But this logic assumes that the selection differential, S, may remain invariant under different heritabilities. The problem then arises from the fact that as $h^2 = V_A/V_P$ (Expression 2.25), low phenotypic variances, V_P, may lead to both higher heritabilities and lower limits to the selection differentials, S, that can be applied in practice (as no or very little individuals will have phenotypes far away from the mean when the phenotypic variance is low). Overall, high heritabilities may not imply large responses to selection in practice because they may impose limitations on the selection differentials, which has actually been verified empirically (Houle 1992).

To avoid that problem, Houle (1992) proposed to build an alternative index using a different denominator, not correlated with the numerator. In particular, he proposed to scale the additive variance with the squared population average, μ^2, (i.e., a mean-standardized scale) instead of with the phenotypic variance. The resulting measure is claimed to be a better measure of evolvability. Indeed, it has a more straightforward biological interpretation as the "percent change in a trait under a unit strength of selection" (see Hansen et al. 2011 and references therein). In relation with this, note that the method proposed in Expression 7.3 to express the variance decomposition with NOIA naturally provides μ^2 as an additional parameter. More in general, NOIA (and particularly ARNOIA) provides expressions of the additive variance under much more complex conditions (genetic architectures and population facts) than any previous models. This is an advantage when studying evolvability under the bottom-up approach (based on information about the genetic basis of the trait), rather than under the top-down approach (based on phenotypes of relatives). As a general take-home message, it seems that approaches to evolvability from a top-down approach should always be considered with caution and double-checked with methods fitting the bottom-up approach (e.g., with NOIA) whenever possible.

10.3.3 Interpret Your Numbers

It has already been stressed several times throughout this book that the components of the genetic variance, particularly the interaction components, are often interpreted loosely. Particularly, it is not always taken into account that even low values of interaction variances can occur under certain circumstances in the face of interactions that play a crucial evolutionary role at a population. This is just one of the conclusions reached when reviewing the interpretation of the numbers of genetic variance decomposition under the heading "Definition and meaning of the variance decomposition" in Chap. 7. By computing the variance decomposition with NOIA, it has been shown that interaction variances in general and the epistatic variance in particular may naturally attain high values relative to the other components of variance under realistic assumptions (Álvarez-Castro and Le Rouzic 2015). These results are relevant particularly when recalling that opposite interpretations were reached using more limited models of variance decomposition (e.g., Hill et al. 2008).

10.3.4 Respect Scale Type

Here there will only be recalled that the models of genetic effects provide quantitative genetics indexes as long as the phenotypes are assigned to numerical values. The components of the genetic variance, for instance, are functions of the population frequencies and the genotypic values. The latter are the expected phenotypes of the different genotypes and for them to be numbers the phenotypes of the individuals must also be so. This is why when we addressed seed color in beans (in Chap. 4), as studied by Mendel, we had to turn from a nominal scale (green, yellow) into a numerical one (zero, one). We decided in particular to use a numerical measure of "yellowness," thus assigning zero to green and one to yellow, but other options could be considered.

Indeed, the trait in question is meristic, with only two numbers necessary to express the phenotypes. Still, the scale type can be considered an interval, enabling to accommodate mutations to intermediate colors or to more extreme ones (like towards red beyond yellow, for instance), thus assuming that the trait is of continuous variation. Incidentally, the quantitative genetics framework naturally leads to quantitative measurements, whether the traits analyzed are of continuous variation or meristic. The type of scale to measure the additive variance of the "yellowness" of beans at a population is an interval (positive real numbers), whether the trait itself is considered to be strictly meristic (with the phenotype in ordinal scale) or not.

10.3.5 Do Not Let Statistics Overrule Meaning

Under this heading, it is equally suitable (as it already was two headings above) to warn about the meaning of the numbers obtained by means of ANOVA as components of the genetic variance. In particular, they are informative about the contribution of the different genetic components—into which the genotypic values were split—to the phenotypic variance at the population in question. Their meaning must not be over-interpreted as providing information about the evolutionary significance of the genetic components in relation to mid- or long-term selection.

There is another message provided in the later chapters of this book (see particularly Chap. 7) that matches this heading. Orthogonality is a statistical property that models of genetic effects must fulfill in order for their population parameters to provide the meaning we expect from them. As mentioned right above, the decomposition of the genetic variance provides the separate contribution of each genetic component to the phenotypic variance, for which an orthogonal model (suitable to analyze the components separately) is necessary. Still, the statistical property does not provide the meaning by itself. In other words, there exist models providing orthogonal decompositions of the genetic variance that do not provide measurements of variance components matching the desired meaning (i.e. the aforementioned biological interpretation).

10.3.6 Treat Measurements as Measurements

Houle et al. (2011) use this token to denounce that wrong estimates of a particular parameter were reported over a few decades just as a particular case of "a systemic lack of respect for measurement and models in biology." Gene mapping remains today one another illustrative example of this problem, to the best of the knowledge of the author of this book. As explained above in this chapter, quantitative trait locus (QTL) analysis in particular and gene mapping in general have been subject to disappointment after several decades of being applied. The optimized models of genetic effects provided in this book bring a breath of fresh air, through the challenge of assessing how much their implementation in mapping tools can improve the results obtained. Beyond that, it can even be interpreted as a paradox that more advanced genetic modeling (specifically, the original setting of NOIA) has been applied to an initially intended top-down line of work (genomic selection) than to the bottom-up concept (gene mapping) that served as the motivation for the improvement of models of genetic effects (see Chap. 9).

The results of QTL analyses are measurements (of relative phenotypic/environmental effects of different variants of loci/environmental conditions) and as such their precise meaning relies on the model (of genetic/environmental effects) used to obtain them, which is often not paid enough attention to. As developed in Chap. 7, models providing measurements (i.e., estimates) with biological meaning in the context of the data analyzed should be used in the mapping procedures, which would be technically beneficial (due to statistical advantages of orthogonality). This is not to say that it is to be regretted that many studies have already been conducted under suboptimal conditions in what regards genetic modeling. In relation with this, it must be considered that although soon after QTL analysis was born, Cheverud and Routman (1995) warned that the models used for decades under a top-down approach are not adequate for the gene-mapping bottom-up approach, the development of the models presented in this book was completed long afterwards—about a quarter of a century later.

It is true that the meaning of the measurements obtained in gene-mapping experiments should have been more adequately tracked, as further discussed below. Even more painfully, it is also true that severe violations of measurement theory have been made, like validating loci mapped using models of genetic effects at the individual level that were not found significant with models at the population level (see Chap. 7). In any case, the main point here is that, whatever the particular subfield of application, a more reliable assessment of the possibilities of gene mapping should be made by applying more advanced genetic modeling in the mapping procedures.

10.3.7 Know What Your Parameters Mean

By pointing out above that a statistical concept as orthogonality, whether a necessary condition of our models must not overrule meaning, we have implicitly assumed that we need to set further conditions to ensure that we accurately know the meaning of our parameters. In order for the meaning of the parameters of the models provided in this book to be more easily tracked, we have started by clarifying the effects of allele substitutions at the individual level (Chap. 4). With this, we have resolved a concern Fisher raised about an inconvenient polarity he noticed in his models, which was actually perpetuated into the twenty-first century (as discussed in Chaps. 2, 4, and 5). Proper measurement (unambiguously defining genetic effects as measures of phenotypic effects of allele substitutions) was key for the polarity to vanish. More to the point, with the models at the individual level in Chap. 4, we also made definitely clear the conceptual and mathematical divide between the meaning of the genetic effects at the individual and at the population levels.

When moving from the individual level to the more complex population level, we have first focused on one-locus cases and then moved to multilocus cases and cases involving environmental factors that can be worked out as a combination of single loci and single environments previously worked out separately (Chap. 5). To end up with, we addressed the most general complexity of models that have to be worked out involving multiple factors (loci and/or environments) from scratch (Chap. 6). We have also made sure that all population models in Chaps. 5 and 6, which were originally developed in different ways, are here developed from the same mathematical scheme (Kempthorne's (1957) regression framework) and in a manner coherent with what had previously worked out at the individual level.

As mentioned in Chap. 6, the individual and the population levels were labeled in different ways, particularly when focusing on epistasis. The individual level was aimed in models called functional, physiological, biological, genetic, or compositional, whereas the population level was commonly addressed under the label statistical (e.g., Cheverud and Routman 1995; Hansen and Wagner 2001; Moore 2005; Moore and Williams 2005; Phillips 2008). NOIA unified both perspectives under the same mathematical framework (Álvarez-Castro and Carlborg 2007), making it crystal clear that the two of them look at the same phenomenon from different angles. Therefore, functional and statistical epistasis are not two different phenomena, but two different ways of analyzing the same phenomenon. The parameters of the formulations at the individual and the population levels have different meanings for matching the perspectives aimed—the former are based on allele substitutions made from individual genotypes, while the latter are based on average effects of them over populations.

We have also previously in this book (Chap. 8) provided a specific application of being aware of the meaning of the parameters of the quantitative genetics models. The "genetic coefficient of the disease" (at a population) is an important index in medical models accounting for interaction effects. Such index stands for the "difference between the additive expectations of case genomes and control genomes" (e.g., Li et al. 2019). We have made use of the fact that in quantitative genetic models of gene-environment interaction, that index fits to twice the additive effect of the genetic component of the model, 2α.

Also the difficulties to conceive that or how an appropriate formulation of genetic effects properly accounting for departures from linkage equilibrium could be achieved can be understood in terms of not being able to account for the precise meaning of the parameters of the models of genetic effects. It was noticed that certain models of genetic effects could account for departures from linkage equilibrium in the absence of other features, like departures from the Hardy-Weinberg proportions or epistasis (e.g., Cockerham 1954; Zeng et al. 2005). But, as already mentioned in Chap. 6, achieving models of genetic effects disentangling departures from linkage equilibrium from any other phenomena has been considered discouraging (e.g., Wang and Zeng 2006; Hill and Mäki-Tanila 2015; Vitezica et al. 2017). Also in Chap. 6, it has been shown that properly implementing arbitrary departures from linkage equilibrium in models of genetic effects with parameters accounting also for whatever other phenomena (which implies in particular completely disentangling population facts from what concerns the genetic architecture) is possible through the correlation-wise orthogonal interactions (COIA) regression framework leading to the ARNOIA upgrade of the NOIA model.

10.3.8 Make Meaningful Measures

The last warning by Houle et al. (2011) is a particularly general one, wherefore it matches a number of relevant features of NOIA. The article providing the first estimates of genetic effects at the individual level from a QTL analysis of real data uses NOIA and has the word "meaningful" in its title (Álvarez-Castro et al. 2008). In statistics, the meaningfulness of several procedures is addressed through the word "coherent" (e.g., Everitt 2002; Dodge 2003). That word is in the title of the article where the interval mapping by imputations (IMI) method enabling the use of NOIA to obtain orthogonal estimates of genetic effects in the face of missing genotype information is published (Nettelblad et al. 2012). We shall start by reviewing the general context that makes those words to be suitable for those titles.

For being really useful, particularly for bottom-up approaches, models of genetic effects must be flexible in being able to fit to different meanings of their parameters. In order to make meaningful measures of genetic effects in a mapping experiment, the first models used have to fit the population frequencies of the different loci as they are analyzed. This is the way that enables to properly extract the most likely genetic architecture from the data. Then, a researcher will typically have in mind a question that motivated the study. It is at this point key to identify the particular model providing the estimates matching the meaning necessary to answer that research question properly. Hereafter, we briefly describe the steps that must be followed for properly analyzing the data and then address the question or questions that motivated the study (see also Álvarez-Castro 2012; Álvarez-Castro et al. 2012, and the last section of Chap. 7).

First, it is necessary to assess each potential underlying locus or group of loci with models that match those portions of the data (rather than matching the question that motivates the study), in order to adequately analyze all the information the data can provide. These models can be built using NOIA at the population level, as described throughout Chaps. 5 and 6. For instance, when analyzing a single biallelic locus, the model comes from Expression 5.11, using the frequencies of the locus in question in the data. As seen in Chap. 7, for the properties of the model not to be misused, also the regression procedure must properly account for missing genotype information—using IMI.

This way, model selection is performed by comparing the fit of the different (groups of) loci, and the most reliable estimates of genetic effects are obtained with models matching the particularities of the data. It is at this point that the estimates fitting the meaning of the question(s) motivating the study can be obtained. If genotypic values are required, they should be obtained by means of genetic filtering, rather than directly from the data. Genetic filtering provides estimates coherent with the meaning of the genetic effects assumed in the study (i.e., the ones found significant in the analyses) and in some cases actually the only estimates possible (see Chap. 7; Álvarez-Castro 2012). Finally, if genetic effects are needed under a reference that differs from the one used to match the data, then the transformation tool must be used accordingly, i.e., with the particular model matching the question to be answered (see Chap. 3).

As discussed in Chap. 9, a completely general model of genetic effects—capable to fit all possible features of the data—was not available in the early days of QTL analysis. It is thus understandable that analyses were performed with the models available, which does not squarely fit measurement theory. Assumingly, the consequences have been both an increase of false positives, an increase of false negatives, and less precise estimates than could be obtained with more advanced genetic modeling. What is not so understandable is to make misuse of the more recent improvement in flexibility of the new developments in models of genetic effects. More in particular, a much worse violation of measurement theory (thus leading to less reliable results) than carrying out gene-mapping with limited models at the population level (that cannot completely fit the population frequencies of the data) is to use models at the individual level, which are just inappropriate for this purpose (e.g., Xiao et al. 2014; Deng et al. 2020).

There have been published bioinformatics resources enabling various aspects of the use of NOIA. Le Rouzic and Álvarez-Castro (2008) released a tool, "noia," to obtain genotypic values from genetic effects and vice-versa and to apply the aforementioned transformation tool, thus enabling genetic filtering and transformations based on the NOIA original setting (by Álvarez-Castro and Carlborg 2007). Also MAPfastR, a tool for QTL analysis of both inbreed and outbreed populations (by Nelson et al. 2013) implemented a version of NOIA for multiple biallelic loci, but it seems to be no longer available. At the time of publication of this book, no user friendly tool implementing NOIA (and particularly ARNOIA) in a way that the steps described above can be applied. Thus, some applications of NOIA in bottom-up approaches involve either programming or contacting the authors of the above bioinformatics tools and asking them to provide further implementations of the already developed tools.

Beyond the genetic effects themselves, Houle et al. (2011) also mention an index of directional epistasis as an illustrative example of a meaningful measurement. This parameter was proposed by Carter et al. (2005) to predict the output of mid- and long-term selection. Although originally defined in the context of a multilinear genetic architecture (Hansen and Wagner 2001), such index can be obtained using NOIA from an arbitrary epistatic genetic architecture (Pavlicev et al. 2010; Le Rouzic 2014). The directional epistasis index can be obtained from gene-mapping experiments—actually, using one of the applications already implemented in a bioinformatics tool, "noia" (Le Rouzic and Álvarez-Castro 2008).

Thus, the index of directional epistasis is defined under the bottom-up approach. Indeed, it illustrates that this approach can provide sounder procedures to address evolvability, particularly in what regards making predictions that are really meaningful beyond a one-generation step, which is one of the important limitations of heritability. In particular, it is has been shown that heritability may substantially vary in only a few generations due to epistasis, even from significant values to negligible ones and vice-versa (e.g., Álvarez-Castro and Le Rouzic 2015; Goodnight 2015; Le Rouzic and Álvarez-Castro 2016; also see above).

10.4 Measurement of Heritability

We here make a list of the concerns about the use of heritability that have been addressed throughout this book. On top of the one mentioned right above (heritability is not a meaningful measure beyond very short-term selection) two others have been discussed. On the one hand, the measurement of heritability from a top-down approach is involved with several theoretical problems that arise particularly due to epistasis, including those discussed in Chaps. 2 and 7, and considered in analyses presented in Chap. 8, and the ones pointed out by Zuk et al. (2012). On the other hand, as discussed above, even when measured in a theoretically correct way (e.g., using a bottom-up approach), it is not meaningful to use heritabilities in order to compare the potential response to selection that can be achieved in different traits—even if we restrict our comparison to a one-generation step.

10.5 Test Passed

In conclusion, NOIA has here been found to align with the principles of measurement theory. En passant, we have reviewed the content of this book from different perspectives. This is to say, this addendum hopefully gives the reader an opportunity to rework the most important concepts provided throughout this book.

References

Álvarez-Castro JM (2012) Current applications of models of genetic effects with interactions across the genome. Curr Genomics 13:163–175

Álvarez-Castro JM, Carlborg Ö (2007) A unified model for functional and statistical epistasis and its application in quantitative trait loci analysis. Genetics 176:1151–1167

Álvarez-Castro JM, Carlborg O, Ronnegard L (2012) Estimation and interpretation of genetic effects with epistasis using the NOIA model. Methods Mol Biol 871:191–204

Álvarez-Castro JM, Le Rouzic A (2015) On the partitioning of genetic variance with epistasis. In: Moore JH, Williams SM (eds) Epistasis: methods and protocols. Springer, Humana Press, New York, pp 95–114

Álvarez-Castro JM, Le Rouzic A, Carlborg Ö (2008) How to perform meaningful estimates of genetic effects. PLoS Genet 4:e1000062

Carter AJ, Hermisson J, Hansen TF (2005) The role of epistatic gene interactions in the response to selection and the evolution of evolvability. Theor Popul Biol 68:179–196

Cheverud JM, Routman EJ (1995) Epistasis and its contribution to genetic variance components. Genetics 139:1455–1461

Cockerham CC (1954) An extension of the concept of partitioning hereditary variance for analysis of covariances among relatives when epistasis is present. Genetics 39:859–882

Deng S, Hardin J, Amos CI, Xiao F (2020) Joint modeling of eQTLs and parent-of-origin effects using an orthogonal framework with RNA-seq data. Hum Genet 139:1107–1117
Dodge Y (2003) The Oxford dictionary of statistical terms. Oxford University Press, Oxford
Everitt BS (2002) The Cambridge dictionary of statistics. Cambridge University Press, Cambridge
Fisher RA (1918) The correlation between relatives on the supposition of Mendelian inheritance. Trans Roy Soc Edinburgh 52:339–433
Fisher RA (1930) The genetical theory of natural selection. Clarendon, Oxford
Fisher RA (1941) Average excess and average effect of a gene substitution. Ann Eugenics 11:53–63
Goodnight C (2015) Long-term selection experiments: epistasis and the response to selection. In: Moore JH, Williams SM (eds) Epistasis. Methods and protocols. Springer, New York, pp 1–18
Hansen TF (2016) Evolvability, quantitative genetics of. In: Kliman RM (ed) Encyclopedia of evolutionary biology. Academic Press, Oxford, pp 83–89
Hansen TF, Pélabon C, Houle D (2011) Heritability is not evolvability. Evol Biol 38:258–277
Hansen TF, Wagner GP (2001) Modeling genetic architecture: a multilinear theory of gene interaction. Theor Popul Biol 59:61–86
Hill WG, Goddard ME, Visscher PM (2008) Data and theory point to mainly additive genetic variance for complex traits. PLoS Genet 4:e1000008
Hill WG, Mäki-Tanila A (2015) Expected influence of linkage disequilibrium on genetic variance caused by dominance and epistasis on quantitative traits. J Anim Breed Genet 132:176–186
Houle D (1992) Comparing evolvability and variability of quantitative traits. Genetics 130:195–204
Houle D, Pelabon C, Wagner GP, Hansen TF (2011) Measurement and meaning in biology. Q Rev Biol 86:3–34
Kempthorne O (1957) An introduction to genetic statistics. Wiley, New York
Krantz DH, Luce RD, Suppes P, Tversky A (1971) Foundations of measurement. Academic Press, New York
Le Rouzic A (2014) Estimating directional epistasis. Front Genet 5:198
Le Rouzic A, Álvarez-Castro JM (2008) Estimation of genetic effects and genotype-phenotype maps. Evol Bioinforma 4:225–235
Le Rouzic A, Álvarez-Castro JM (2016) Epistasis-induced evolutionary plateaus in selection responses. Am Nat 188:E134–E150
Levins R (1966) Strategy of model building in population biology. Am Sci 54:421–431
Lewontin RC (1974) The genetic basis of evolutionary change. Columbia University Press, New York
Li J, Li X, Zhang S, Snyder M (2019) Gene-environment interaction in the era of precision medicine. Cell 177:38–44
Moore JH (2005) A global view of epistasis. Nat Genet 37:13–14
Moore JH, Williams SM (2005) Traversing the conceptual difice between biological and statistical epistasis: systems biology and a more modern synthesis. ByoEssays 27:637–646
Nelson RM, Nettelblad C, Pettersson ME, Shen X, Crooks L, Besnier F, Alvarez-Castro JM, Ronnegard L, Ek W, Sheng Z, Kierczak M, Holmgren S, Carlborg O (2013) MAPfastR: quantitative trait loci mapping in outbred line crosses. G3 3:2147–2149
Nettelblad C, Carlborg Ö, Pino-Querido A, Álvarez-Castro JM (2012) Coherent estimates of genetic effects with missing information. Open J Genet 2:8
Pavlicev M, Le Rouzic A, Cheverud JM, Wagner GP, Hansen TF (2010) Directionality of epistasis in a murine intercross population. Genetics 185:1489–1505
Phillips PC (2008) Epistasis—the essential role of gene interactions in the structure and evolution of genetic systems. Nat Rev Genet 9:855–867
Russo L (2004) The forgotten revolution: how science sas born in 300 BC and why it had to be reborn. Springer, Heidelberg
Vitezica ZG, Legarra A, Toro MA, Varona L (2017) Orthogonal estimates of variances for additive, dominance, and epistatic effects in populations. Genetics 206:1297–1307
Wang T, Zeng ZB (2006) Models and partition of variance for quantitative trait loci with epistasis and linkage disequilibrium. BMC Genet 7:9

Xiao F, Ma J, Cai G, Fang S, Lee JE, Wei Q, Amos CI (2014) Natural and orthogonal model for estimating gene-gene interactions applied to cutaneous melanoma. Hum Genet 133:559–574
Zeng ZB, Wang T, Zou W (2005) Modeling quantitative trait loci and interpretation of models. Genetics 169:1711–1725
Zuk O, Hechter E, Sunyaev SR, Lander ES (2012) The mystery of missing heritability: genetic interactions create phantom heritability. Proc Natl Acad Sci U S A 109:1193–1198

Index

J. M. Álvarez-Castro, *Genes, Environments and Interactions*,
https://doi.org/10.1007/978-3-031-41159-5